Process Development Symposium 2018

Applying New Technologies in Process Development

Oak Brook, Illinois, USA
5 - 7 June 2018

ISBN: 978-1-5108-7280-6

Printed from e-media with permission by:

Curran Associates, Inc.
57 Morehouse Lane
Red Hook, NY 12571

Some format issues inherent in the e-media version may also appear in this print version.

Printed by Curran Associates, Inc. (2018)

For permission requests, please contact AIChE
at the address below.

AIChE
120 Wall Street, FL 23
New York, NY 10005-4020

Phone: (800) 242-4363
Fax: (203) 775-5177

www.aiche.org

Additional copies of this publication are available from:

Curran Associates, Inc.
57 Morehouse Lane
Red Hook, NY 12571 USA
Phone: 845-758-0400
Fax: 845-758-2633
Email: curran@proceedings.com
Web: www.proceedings.com

Process Development Symposium 2018

Applying New Technologies in Process Development

Oak Brook, Illinois, USA
5 - 7 June 2018

AIChE
The Global Home of Chemical Engineers

TABLE OF CONTENTS

PROCESS INTENSIFICATION

MODULAR SYSTEMS

CHALLENGES AND OPPORTUNITIES FOR PROCESS DEVELOPMENT

POSTER RECEPTION

3M Science. Applied to Life.™

3M Science: Combining 46 core technologies to improve daily life

2018 AIChE Process Development Symposium

Cristina U Thomas, PhD
Sr. Technical Leader,
Corp R&D Services Center

June 6, 2018

Outline

- Inventor, Commercialization Manager and Senior Technical Leader
- 3M Corporate Overview
- 3M Technologies
- 3M Products and Processes
- 3M Corporate Research Process Laboratory
- State of Science Index
- In Summary

3M

Inventor, Commercialization Manager and Senior Technical Leader

Inventor, Commercialization Manager and Senior Technical Leader

3M™ Raised Pavement Markers Series 290

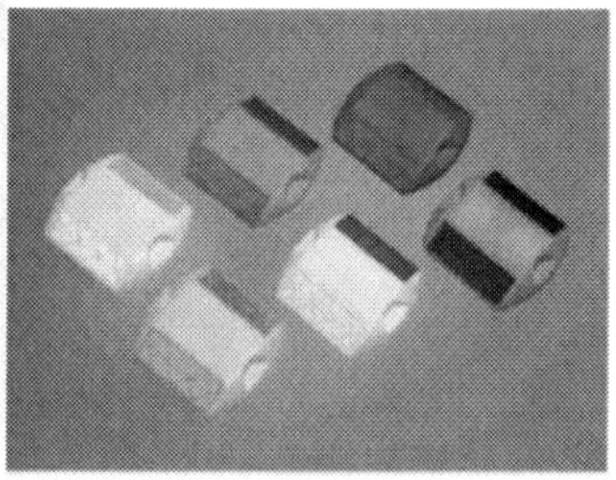

3M™ Stamark High Performance Tape Series

Inventor, Commercialization Manager and Senior Technical Leader

3M™ Sun Control Window Films, Prestige Series for Residential and Commercial

3M™ Automotive Window Film Crystalline Series

3M Corporate Overview

6

Our Vision

3M Technology Advancing Every Company
3M Products Enhancing Every Home
3M Innovation Improving Every Life

3M at a glance

- Sales in ~200 countries
- $32 billion in sales
- Five business groups
- 90,000 3Mers globally
- 113,000 patents
- 101 straight years of dividends
- One of 30 companies on the Dow Jones Industrial Index

Investing in Innovation
3M innovation creating value for our customers

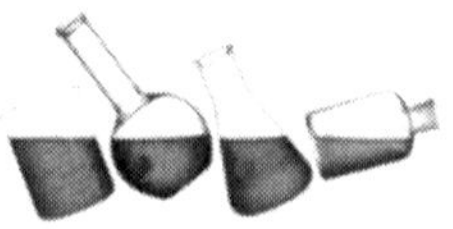

Investing in Innovation

- Enhance commercialization effectiveness
- Invest in and scale new growth platforms
- Increase R&D productivity

3M

3M Technologies

3M

46 Technology Platforms

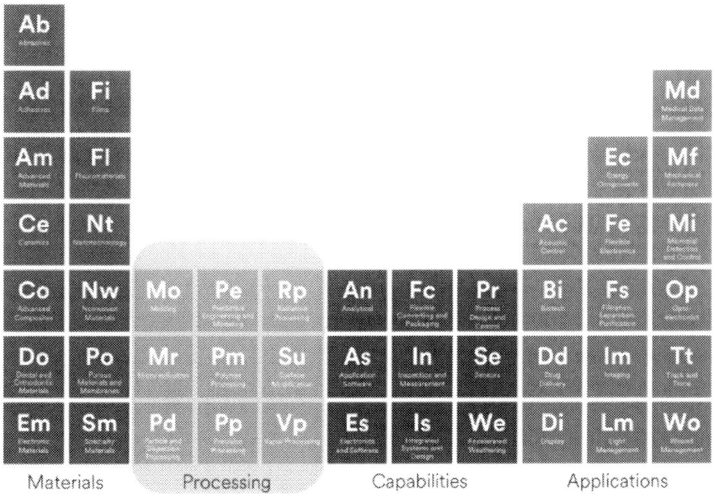

3M

How do we apply science to solve problems and improve lives?

Our people and collaborative culture

\+

Insights from customers and the market

\+

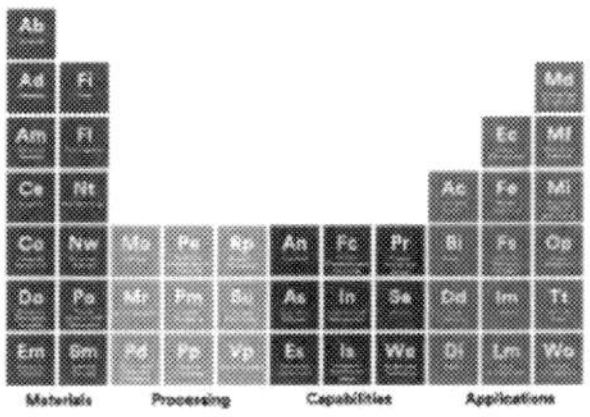

46 technology platforms

3M

Global capabilities

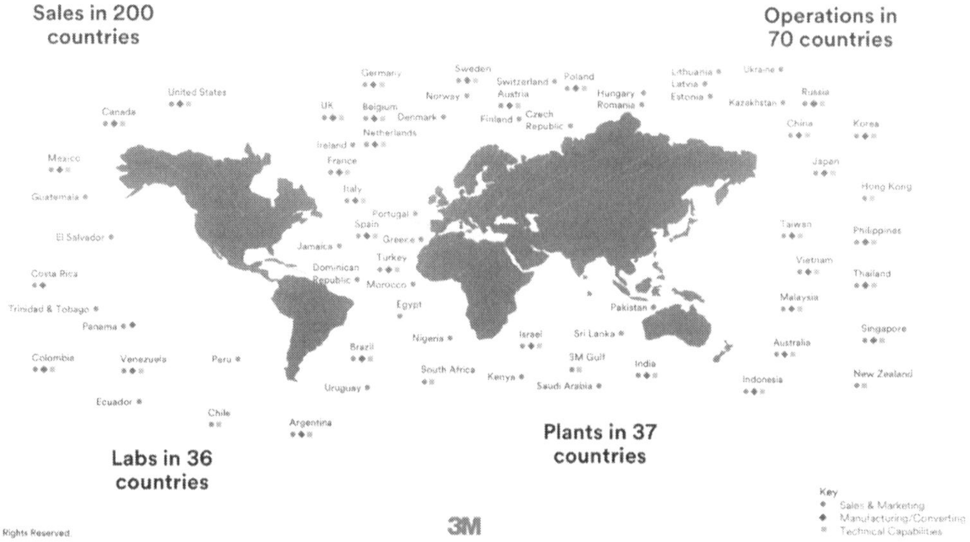

3M

13

3M innovation model

Application of science

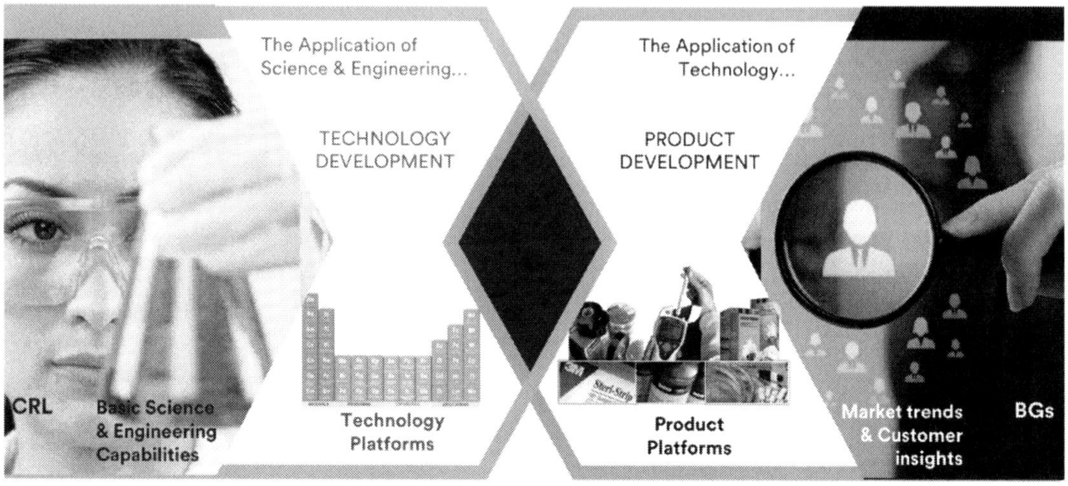

3M

14

Anticipating & preparing for the future

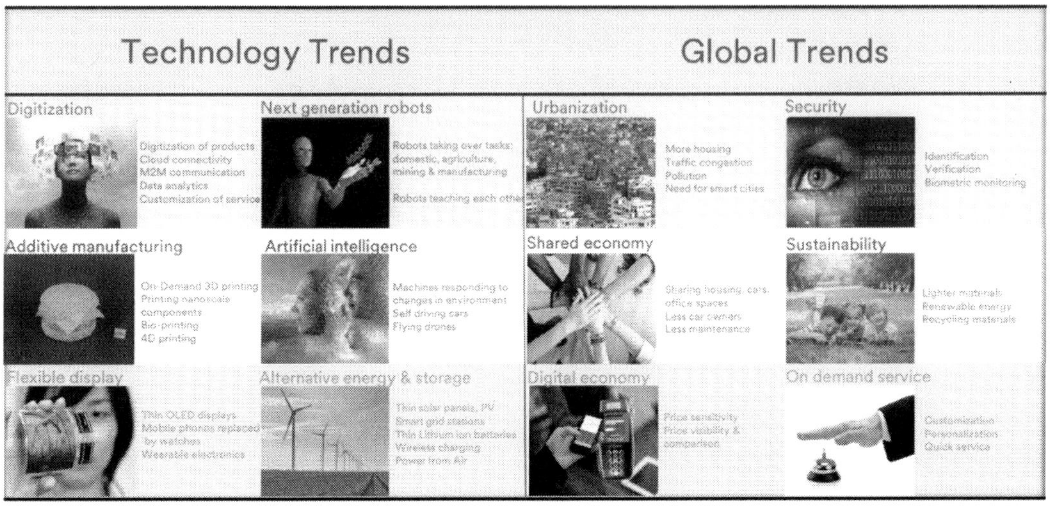

3M

15

3M Corporate Research

interdisciplinary by design

3M Uniqueness:
the intersections and combinations of our technologies

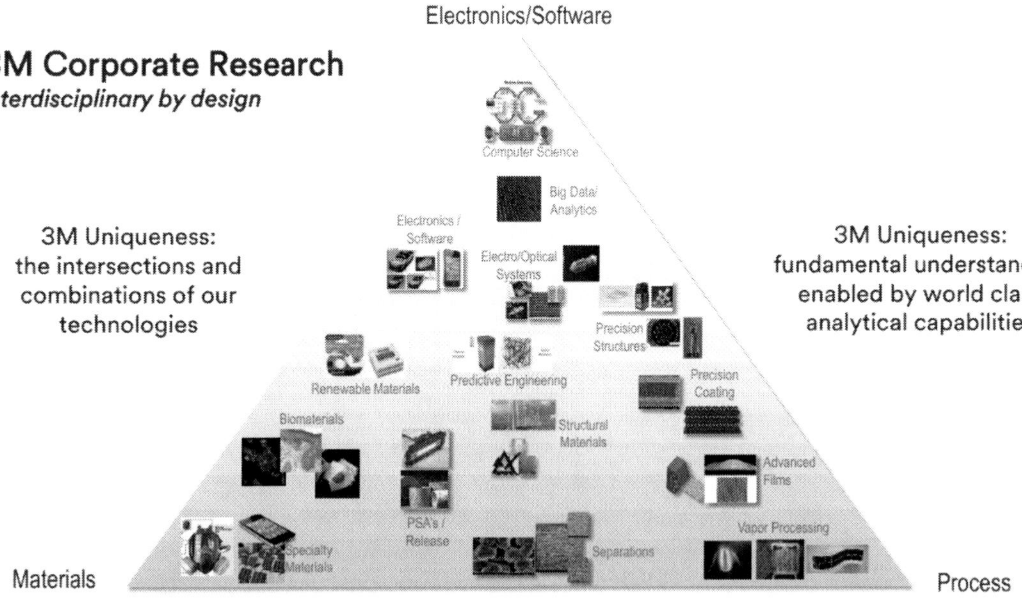

3M Uniqueness:
fundamental understanding enabled by world class analytical capabilities

3M

16

Development & commercialization methodologies

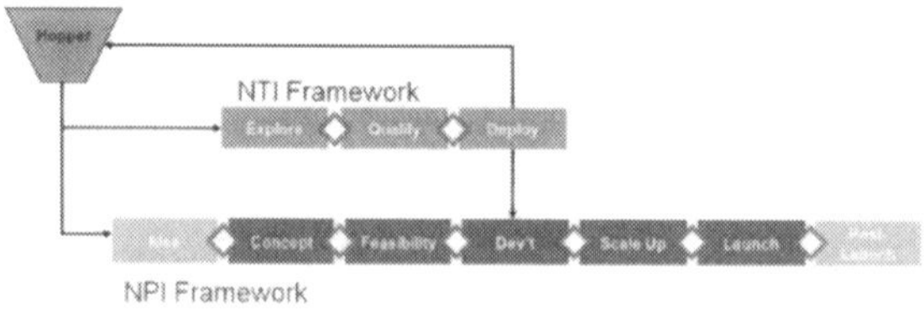

- NTI - To optimize the process of technology identification, development, and successful commercialization.
- NPI - To promote an efficient, standardized commercialization process that supports data driven gate reviews for all businesses worldwide.

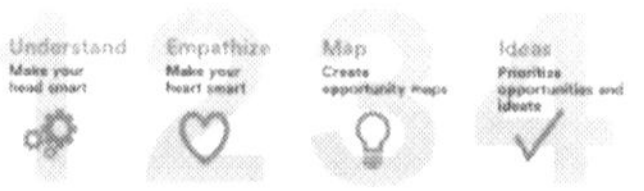

Customer engagement throughout the entire process

3M

A globally connected collaboration community

TECH FORUM

- 12,000+ Members (Grass Roots)
- Annual Event to showcase new products
- Annual Event to highlight new technologies
- Multiple Recognition Events
- Hundreds of Chapter Events

genesis

- Grant Process to stimulate the development of new technologies
- Semi-annual grant process
- Grants can exceed $100K per project
- Multiple Awareness building events

culture 15%

- Mindset to offer permission to work on new ideas that an individual thinks could be a successful product or solution
- Individual initiative but also encourages the collaboration by working with others.

3M

15 percent culture

Precision-shaped grain abrasives

3M Products and Processes

We combine our core technologies in unique and creative ways

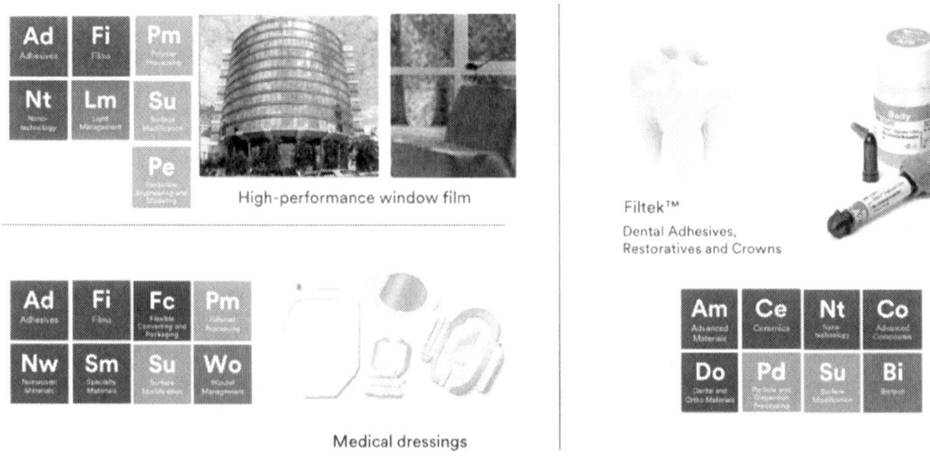

Migration of Microreplication Technology

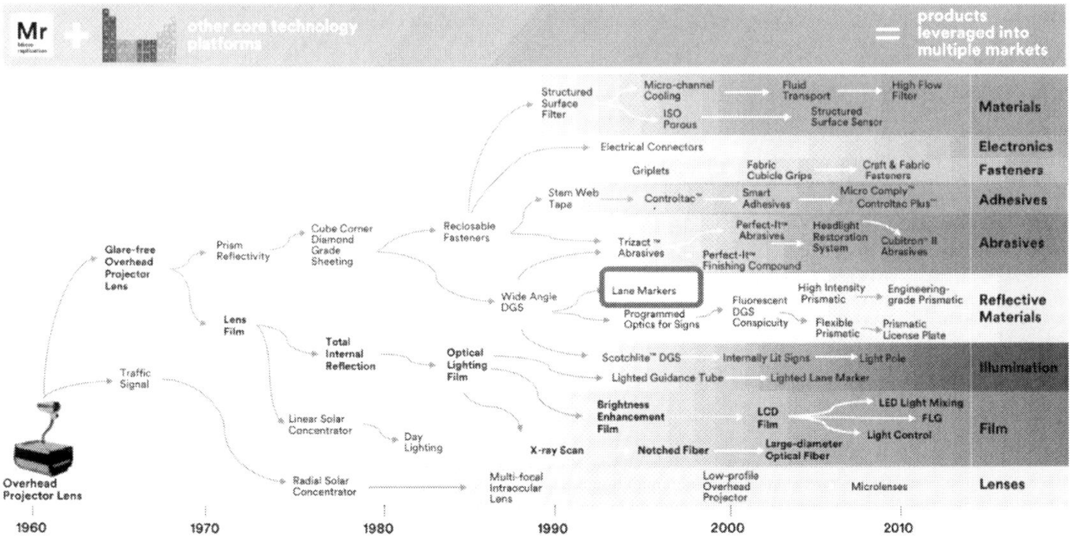

Helping solve sustainability challenges

Increasing energy efficiency

Strengthening renewable energy

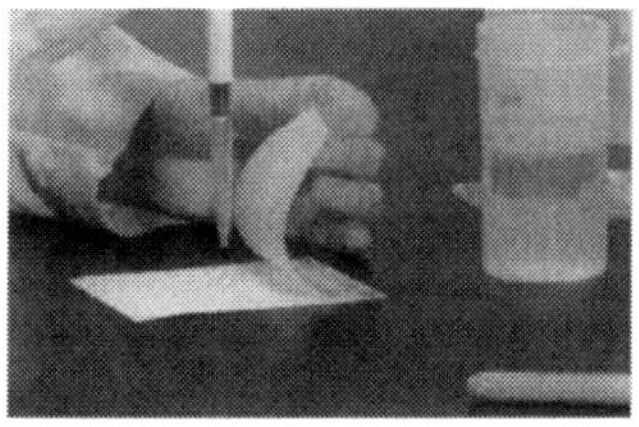
Protecting food supply

Delivering energy where it's needed

Helping people breathe easier

Purifying water

3M

3M Corporate Research Process Laboratory

3M

Corporate Research Process Laboratory

Technology focus areas

3D Processing & Precision Replication

- Additive Manufacturing
- Molding
- Precision Replication

Advanced Materials & Radiation Processing

- Radiation Processing
- Robotics
- Chemical Processing
- Ceramic & Particle Processing
- Thin Film Processing

Specialty Film & Polymer Processing

- Specialty Film Processing
- Polymer Processing
- Porous & Structured Film
- Surface Modification

Precision Coating & Web Processing

- Coating & Modeling
- Printing & Patterning
- Drying & Web Handling

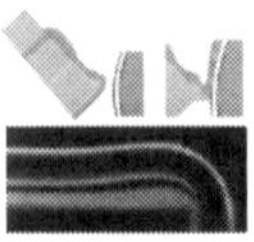

Nonwovens & Ultrasonics

- Nonwovens
- Ultrasonics

Measurement & Process Analytics

- Inspection & Measurement
- Process Analytics

3M

25

......creating technical solutions to drive growth

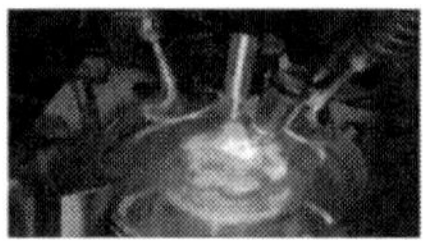

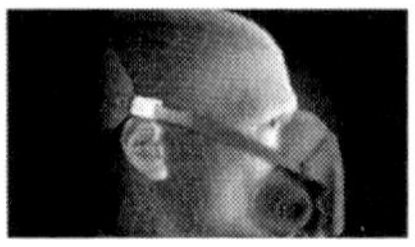

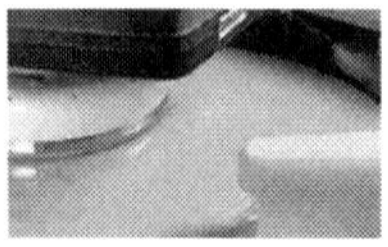

3D Printing

Platform development for Additive Manufacturing (AM)

➢ **Process Technology:**

- Converts digital 3D design to individual components using vat polymerization, binder jetting, or laser sintering

Complex Microstructures

Micro- & nano-fabrication of microfeatures

➢ **Process Technology:**

- Fabrication of complex 3D structures with sub-micron resolution

Molding

Forming, Reactive, Multi-shot, Thermoplastic molding

➢ **Process Technology:**

- Providing precision molding controls, materials, and processing knowledge with precision engineered tooling

Precision Replication

High fidelity macro/micro/nanoreplicated films

➢ **Process Technology:**

- Precision replication and web handling controls to enable process optimization for continuous and discrete microfeatures

3M

26

Differentiated products from state-of-the-art processes

- 3M™ Trizact™ CMP Pad
- 3M™ Trizact™ Pad Conditioner
- 3M™ Cubitron™ Abrasives
- Vikuiti™ Advanced Light Control Film
- 3M™ Reusable Respirators
- 3M™ Microneedle Drug Delivery Systems
- 3M™ Interconnect Solutions
- 3M™ Protective Eyewear
- Automotive Fuel Injector
- 3D Printing of Fully Fluorinated Polymers
- Superabrasives
- Clear Tray Orthodontic Aligner

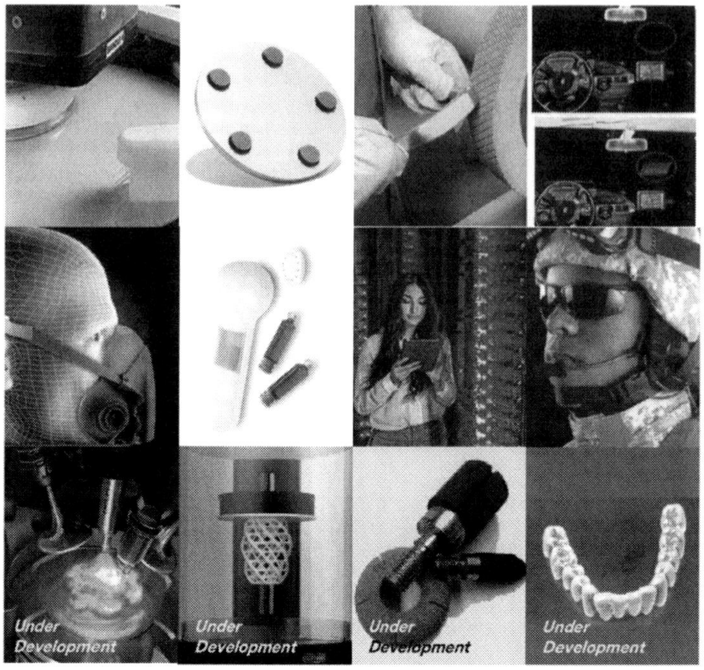

... creating technical solutions to drive growth

Specialty Film

Multi-functional, high performance films via differentiated process technologies

➢ Process Technology:
- Polymer melts are extruded or co-extruded through dies to form flat and structured films

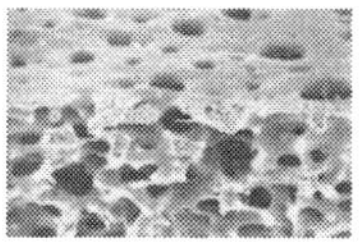

Membranes

Driving growth through engineered porous films and fibers

➢ Process Technology:
- Controlled membrane morphology through phase separation to create precise structures capable of separations

Polymer Processing

Developing innovative and optimized extrusion processing for films and tapes

➢ Process Technology:
- Polymer melts processed through extruders and dies to form films, tapes, and other products of interest

Solventless Adhesives

Significant and differentiated processing of 100% solids adhesives

➢ Process Technology:
- Solventless processing of high performance adhesives (shock absorbing, low VOC, optical clarity, etc.)

 3M 28

Differentiated products from state-of-the-art processes

- 3M™ Kind Removal Silicone Tape
- Scotchgard™ Paint Protection Film
- 3M™ Command™ Strips
- 3M™ Display Enhancement Films
- 3M™ Emphaze™ AEX Hybrid Purifier
- 3M™ Scotchlite™ Reflective Material
- Scotch® Magic™ Tape
- 3M™ Condensation Management Film
- 3M™ Scotchcal™ Graphic Film
- 3M™ VHB™ Tape
- 3M™ Electrical Tape
- 3M™ Prestige Series Window Films

State of Science Index

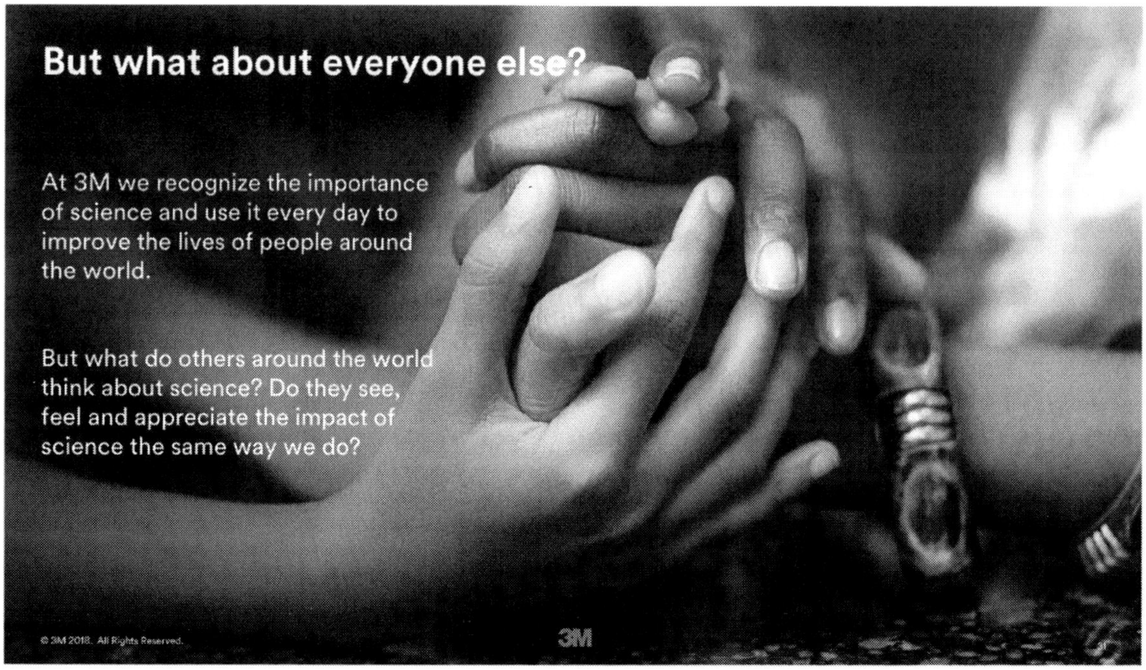

Where we surveyed

Developed ■

Canada, France, Germany, Japan, Singapore, UK, US

Emerging ■

Brazil, China, India, Mexico, Poland, Saudi Arabia, South Africa

- **~1000 participants per country**
- **At the 95% confidence level**
  - 14-country total: +/- 0.83 percentage points
  - Each individual country: +/- 3.10 percentage points

*Based on the MSCI Market Classification 2017: https://www.msci.com/market-classification

3M

32

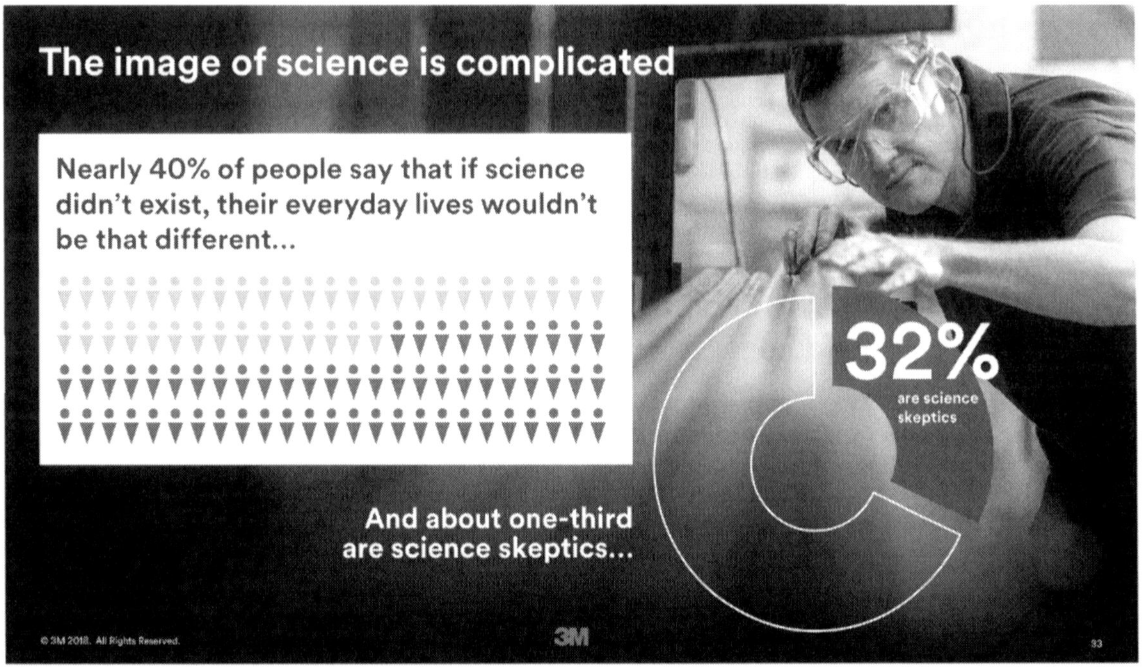

Skeptics have far more negative views

Perceiving science as boring, inaccessible, and unimportant to both society and their everyday life.

When asked if they agree with the following statements, skeptics and non-skeptics said:

(completely or somewhat agree)

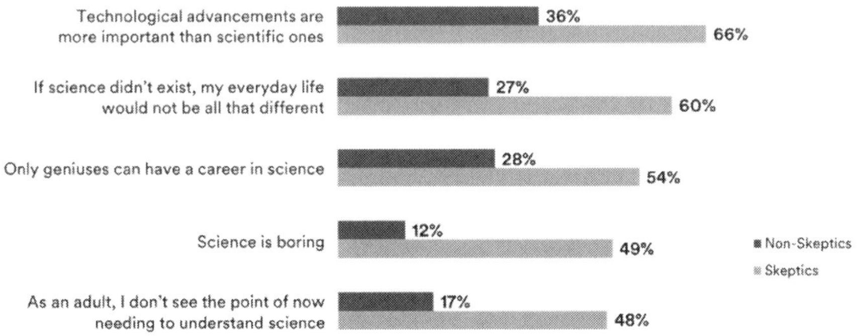

Percentage of respondents who selected somewhat agree or completely agree
Q9: How much do you agree with the following statements?; Q12: How much do you agree or disagree with the following statements? ; Q15: How much do you agree or disagree with the following statements? Skeptics n=4,478; Non-Skeptics n=9,560

3M

34

In Summary: 46 Technology Platforms

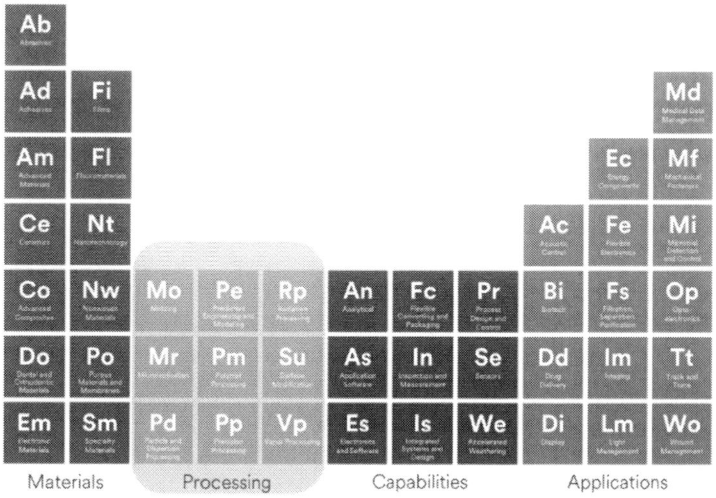

3M

35

azbil

Azbil's Experience Comparing modeling approaches to Optimize a District Cooling Facility

AICHE PDS CONFERENCE JUNE 2018

DAVID HILL — CHEMSTATION
KEN MATSUOKA — AZBIL
MIKE SUZUKI — AZBIL

Chemstations Engineering Advanced 30

azbil

Thirty years ago...

"There is no question that effective optimization algorithms can be meshed with flowsheeting programs . . . "

- T. Edgar and D. Himmelblau,
***Optimization of Chemical Processes,* 1988**

Chemstations Engineering Advanced 30 2

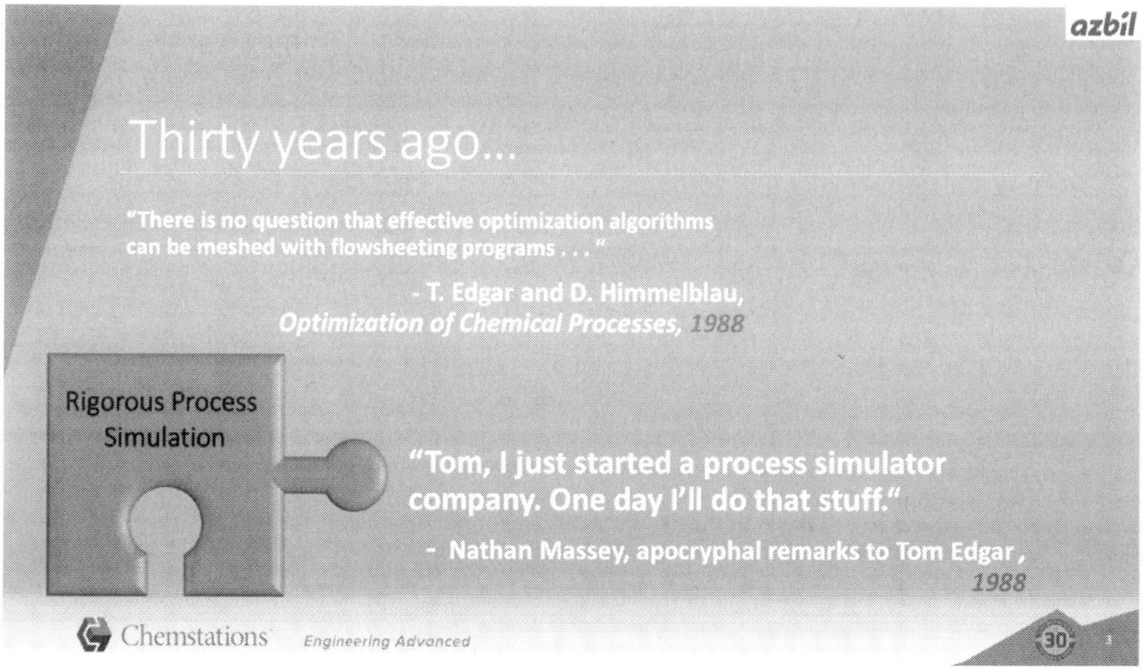
azbil
Thirty years ago...
"There is no question that effective optimization algorithms can be meshed with flowsheeting programs . . . "
- T. Edgar and D. Himmelblau,
Optimization of Chemical Processes, 1988
Rigorous Process Simulation
"Tom, I just started a process simulator company. One day I'll do that stuff."
- Nathan Massey, apocryphal remarks to Tom Edgar , 1988
Chemstations
Engineering Advanced
30

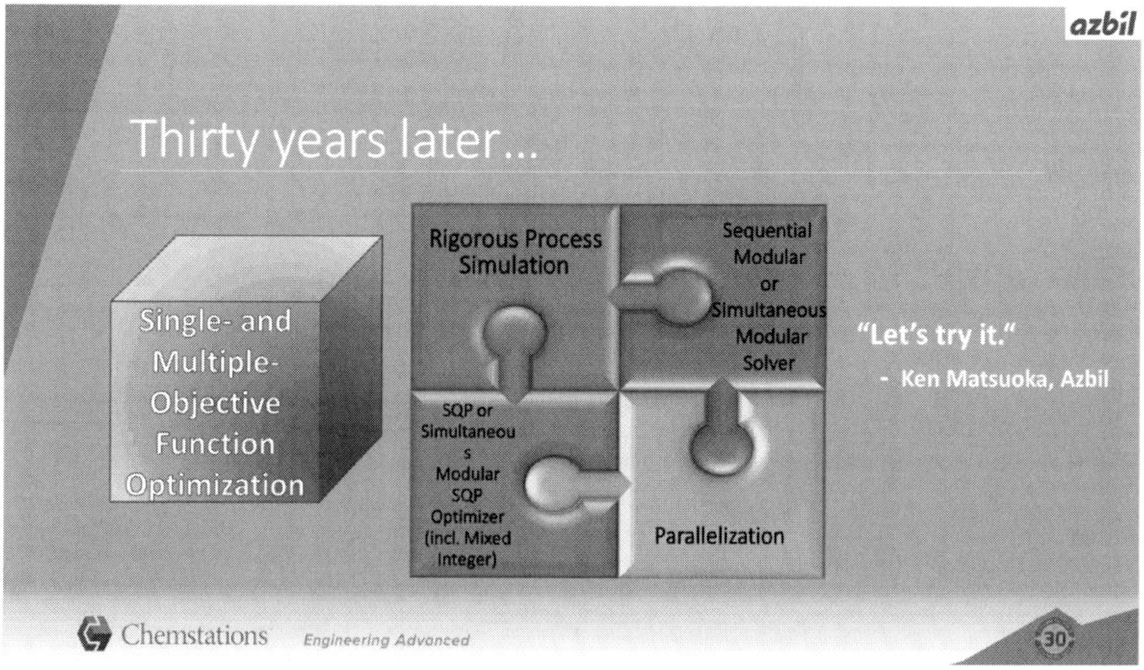
azbil
Thirty years later...
Single- and Multiple-Objective Function Optimization
Rigorous Process Simulation
Sequential Modular or Simultaneous Modular Solver
SQP or Simultaneous Modular SQP Optimizer (incl. Mixed Integer)
Parallelization
"Let's try it."
- Ken Matsuoka, Azbil
Chemstations
Engineering Advanced
30

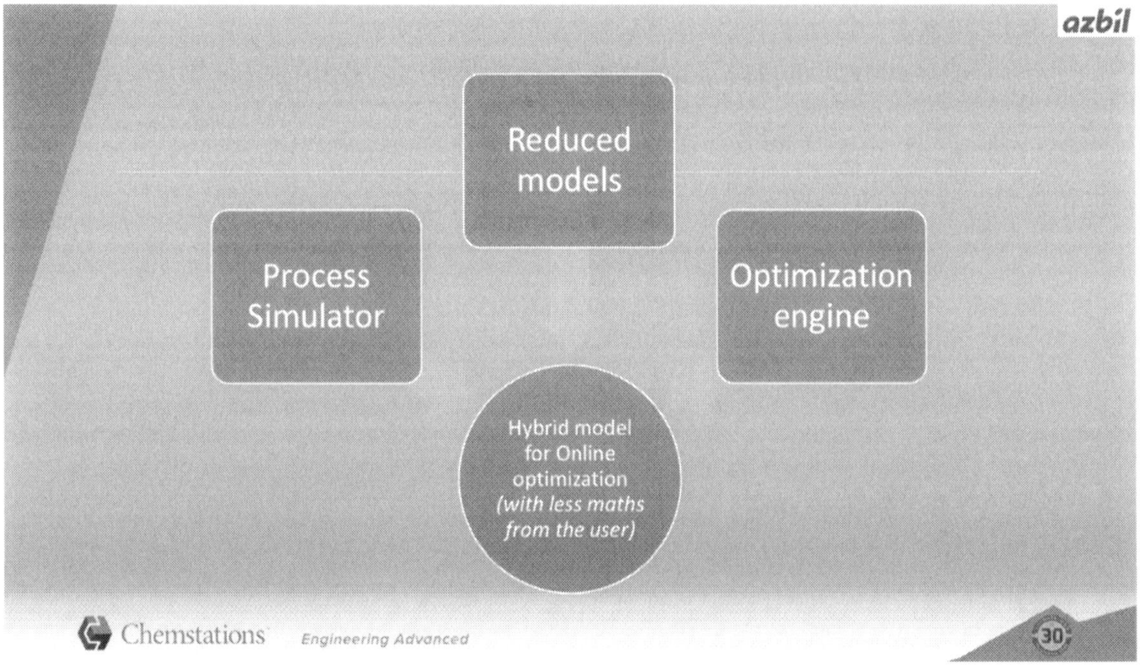

Agenda

- Description of the District Cooling process and goals
- Modeling approach
- Hybrid model for this process – process simulator with reduced models
- Sensitivity studies and optimizations
- Simulator optimization vs. LP optimizer

azbil

District Cooling

Chemstations Engineering Advanced

azbil

District Cooling – chilled water process

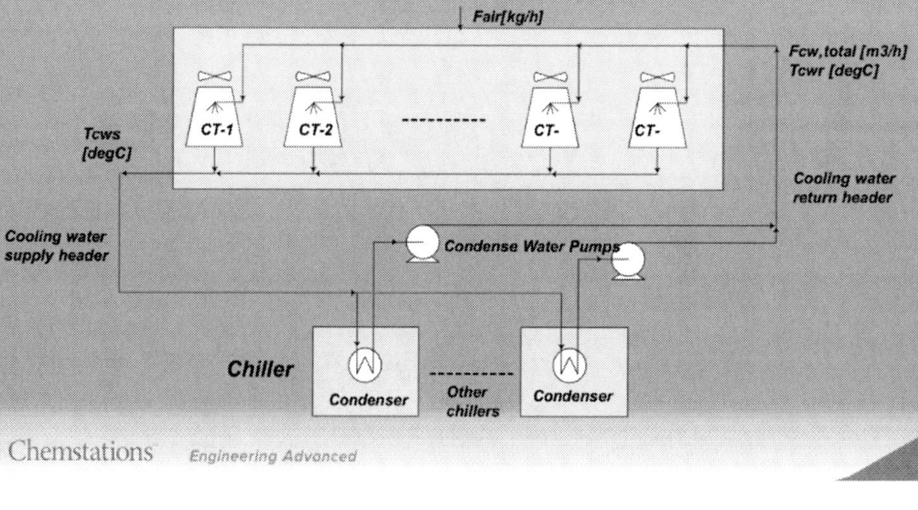

Chemstations Engineering Advanced

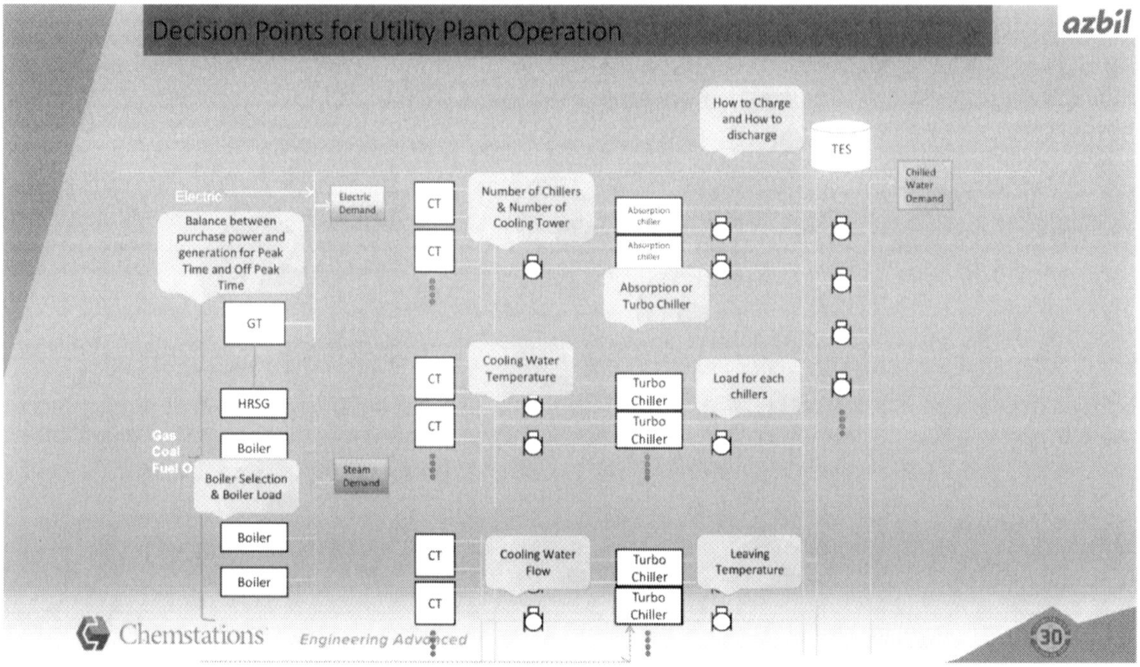
Decision Points for Utility Plant Operation
azbil
How to Charge and How to discharge
TES
Chilled Water Demand
Electric
Electric Demand
Balance between purchase power and generation for Peak Time and Off Peak Time
Number of Chillers & Number of Cooling Tower
Absorption chiller
Absorption or Turbo Chiller
GT
HRSG
Boiler
Boiler Selection & Boiler Load
Steam Demand
Cooling Water Temperature
Load for each chillers
Turbo Chiller
Cooling Water Flow
Leaving Temperature
CT
Chemstations
Engineering Advanced

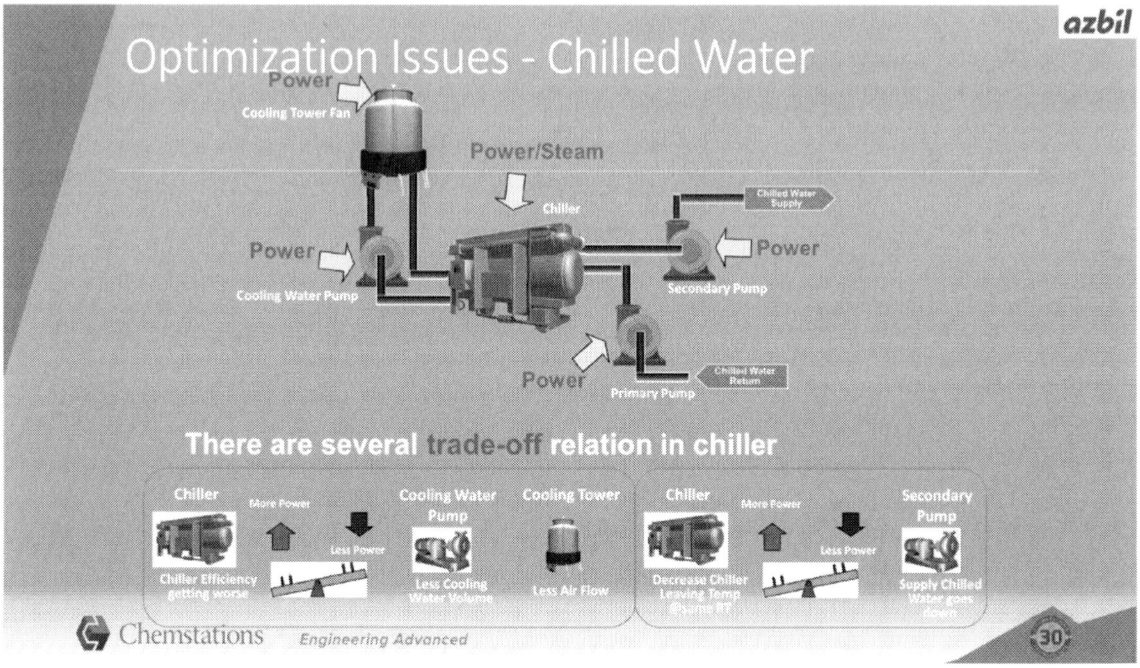
Optimization Issues - Chilled Water
azbil
Power
Cooling Tower Fan
Power/Steam
Chiller
Chilled Water Supply
Power
Cooling Water Pump
Secondary Pump
Power
Primary Pump
Chilled Water Return
There are several trade-off relation in chiller
Chiller
More Power
Less Power
Chiller Efficiency getting worse
Cooling Water Pump
Less Cooling Water Volume
Cooling Tower
Less Air Flow
Chiller
More Power
Less Power
Decrease Chiller Leaving Temp @same RT
Secondary Pump
Supply Chilled Water goes down
Chemstations
Engineering Advanced

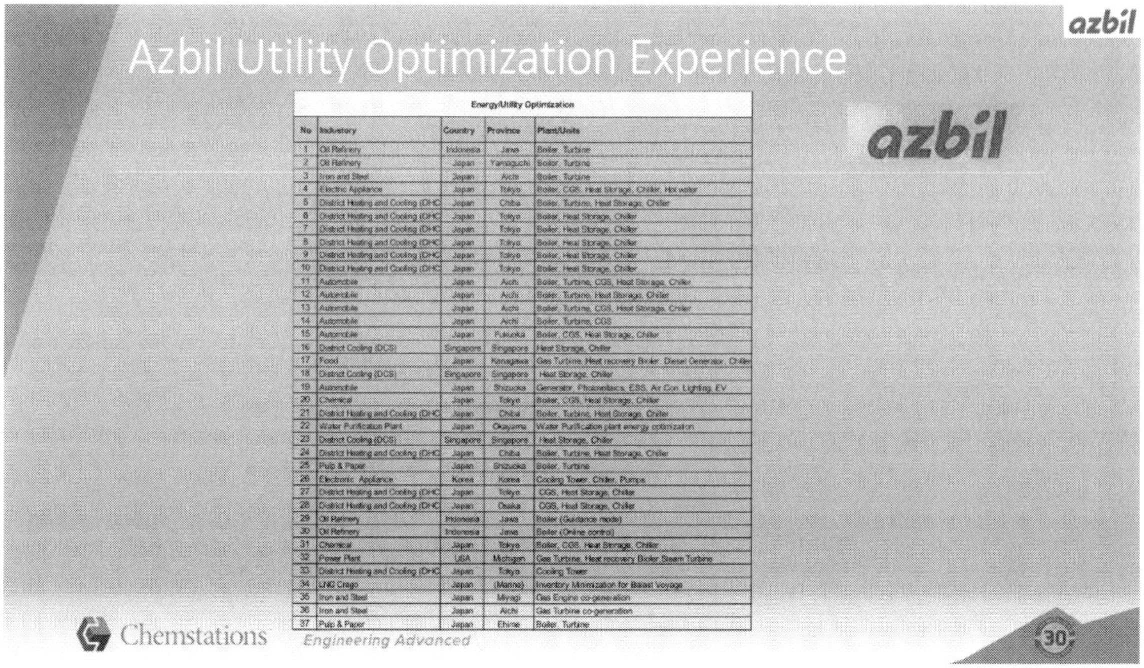

Energy/Utility Optimization

No	Industory	Country	Province	Plant/Units
1	Oil Refinery	Indonesia	Java	Boiler, Turbine
2	Oil Refinery	Japan	Yamaguchi	Boiler, Turbine
3	Iron and Steel	Japan	Aichi	Boiler, Turbine
4	Electric Appliance	Japan	Tokyo	Boiler, CGS, Heat Storage, Chiller, Hot water
5	District Heating and Cooling (DHC	Japan	Chiba	Boiler, Turbine, Heat Storage, Chiller
6	District Heating and Cooling (DHC	Japan	Tokyo	Boiler, Heat Storage, Chiller
7	District Heating and Cooling (DHC	Japan	Tokyo	Boiler, Heat Storage, Chiller
8	District Heating and Cooling (DHC	Japan	Tokyo	Boiler, Heat Storage, Chiller
9	District Heating and Cooling (DHC	Japan	Tokyo	Boiler, Heat Storage, Chiller
10	District Heating and Cooling (DHC	Japan	Tokyo	Boiler, Heat Storage, Chiller
11	Automobile	Japan	Aichi	Boiler, Turbine, CGS, Heat Storage, Chiller
12	Automobile	Japan	Aichi	Boiler, Turbine, Heat Storage, Chiller
13	Automobile	Japan	Aichi	Boiler, Turbine, CGS, Heat Storage, Chiller
14	Automobile	Japan	Aichi	Boiler, Turbine, CGS
15	Automobile	Japan	Fukuoka	Boiler, CGS, Heat Storage, Chiller
16	District Cooling (DCS)	Singapore	Singapore	Heat Storage, Chiller
17	Food	Japan	Kanagawa	Gas Turbine, Heat recovery Boiler, Diesel Generator, Chiller
18	District Cooling (DCS)	Singapore	Singapore	Heat Storage, Chiller
19	Automobile	Japan	Shizuoka	Generator, Photovoltaics, ESS, Air Con, Lighting, EV
20	Chemical	Japan	Tokyo	Boiler, CGS, Heat Storage, Chiller
21	District Heating and Cooling (DHC	Japan	Chiba	Boiler, Turbine, Heat Storage, Chiller
22	Water Purification Plant	Japan	Okayama	Water Purification plant energy optimization
23	District Cooling (DCS)	Singapore	Singapore	Heat Storage, Chiller
24	District Heating and Cooling (DHC	Japan	Chiba	Boiler, Turbine, Heat Storage, Chiller
25	Pulp & Paper	Japan	Shizuoka	Boiler, Turbine
26	Electronic Appliance	Korea	Korea	Cooling Tower, Chiller, Pumps
27	District Heating and Cooling (DHC	Japan	Tokyo	CGS, Heat Storage, Chiller
28	District Heating and Cooling (DHC	Japan	Osaka	CGS, Heat Storage, Chiller
29	Oil Refinery	Indonesia	Java	Boiler (Guidance mode)
30	Oil Refinery	Indonesia	Java	Boiler (Online control)
31	Chemical	Japan	Tokyo	Boiler, CGS, Heat Storage, Chiller
32	Power Plant	USA	Michigan	Gas Turbine, Heat recovery Boiler, Steam Turbine
33	District Heating and Cooling (DHC	Japan	Tokyo	Cooling Tower
34	LNG Crago	Japan	(Marine)	Inventory Minimization for Ballast Voyage
35	Iron and Steel	Japan	Miyagi	Gas Engine co-generation
36	Iron and Steel	Japan	Aichi	Gas Turbine co-generation
37	Pulp & Paper	Japan	Ehime	Boiler, Turbine

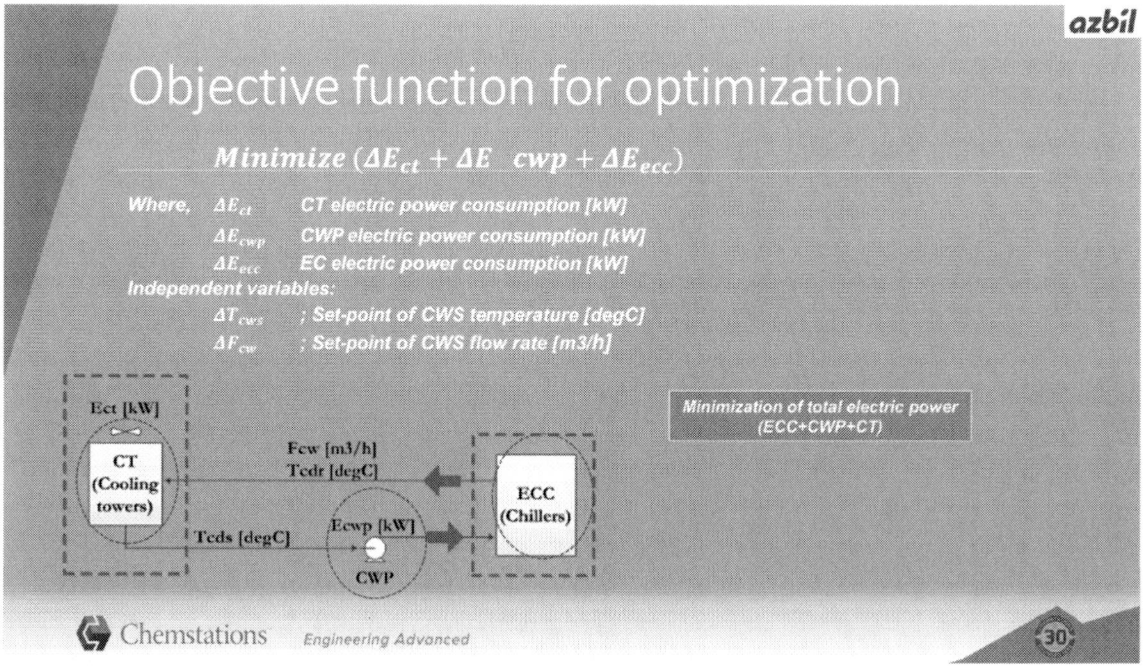

LP approach – reduced model for cooling tower

$\therefore T_{out} = f(F_{cw}, F_{in}, F_{air}) \cong f(F_{cw}, F_{in}, Hz)$

$$\therefore T_{out} = \left(\frac{\partial f}{\partial F_{cw}}\right) \cdot \Delta F_{cw}$$

$$+ \left(\frac{\partial f}{\partial T_{in}}\right) \cdot \Delta T_{in}$$

$$+ \left(\frac{\partial f}{\partial Hz}\right) \cdot \Delta Hz$$

$$k_1 \equiv \left(\frac{\partial f}{\partial F_{cw}}\right)$$

$$k_2 \equiv \left(\frac{\partial f}{\partial T_{IN}}\right)$$

$$k_3 \equiv \left(\frac{\partial f}{\partial Hz}\right)$$

$\therefore T_{out} \cong k_1 \cdot \Delta F_{cw} + k_2 \cdot \Delta T_{in} + k_3 \cdot \Delta Hz$

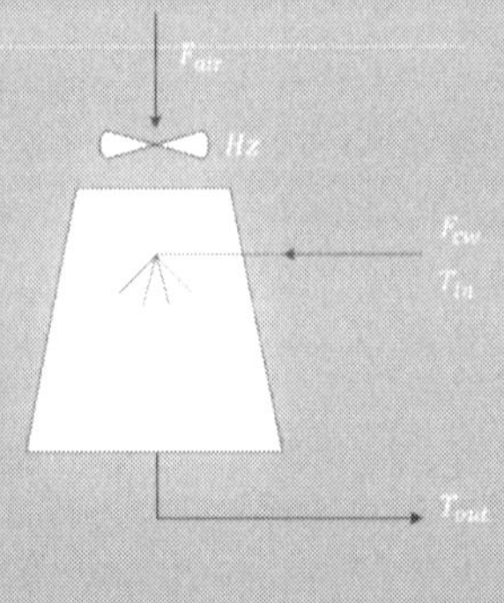

$\Delta F_{cw} = F_{cw,new} - F_{cw,base}$

$\Delta T_{in} = T_{in,wet} - T_{in,base}$

$k_1 = g_1(F_{cw}, F_{in}, Hz)$

Problem of LP / regression Approach

Multiple regression to estimate Cooling water temperature is not accurate

=> Measurement has some error

=> Could not get good correlation

=> Difficult to ask so many step test

=> No consideration for wet bulb temperature (The following diagram is showing actual behavior of approach between cooling water temp and Wet Bulb Temp)

=> DEVELOPMENT TIME IS LARGE

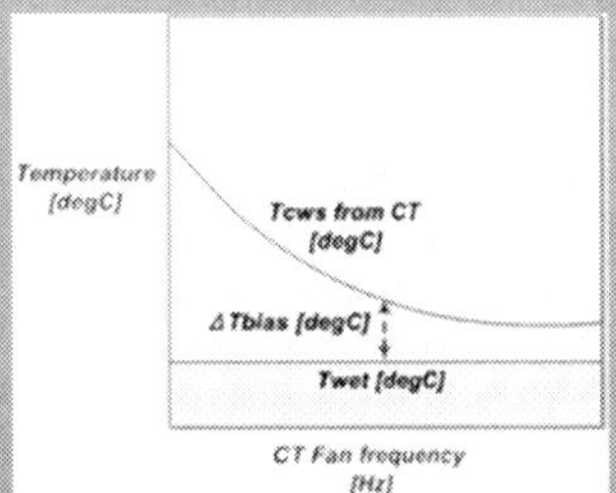

Tcws ; Cooling water temperature leaving from cooling tower [degC]

Twet ; Wet bulb temperature [degC]

ΔTbias; Difference between Tcws and Twet [degC]

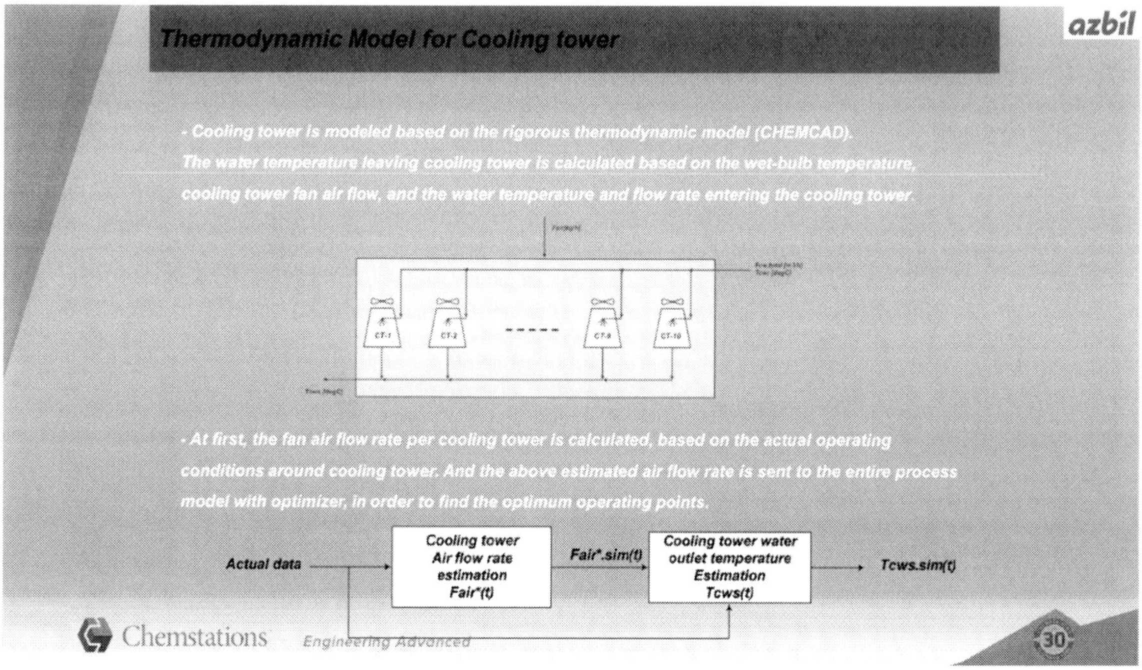
Thermodynamic Model for Cooling tower
azbil
- Cooling tower is modeled based on the rigorous thermodynamic model (CHEMCAD).
The water temperature leaving cooling tower is calculated based on the wet-bulb temperature, cooling tower fan air flow, and the water temperature and flow rate entering the cooling tower.
- At first, the fan air flow rate per cooling tower is calculated, based on the actual operating conditions around cooling tower. And the above estimated air flow rate is sent to the entire process model with optimizer, in order to find the optimum operating points.
Actual data
Cooling tower Air flow rate estimation Fair*(t)
Fair*.sim(t)
Cooling tower water outlet temperature Estimation Tcws(t)
Tcws.sim(t)
Chemstations
Engineering Advanced
30

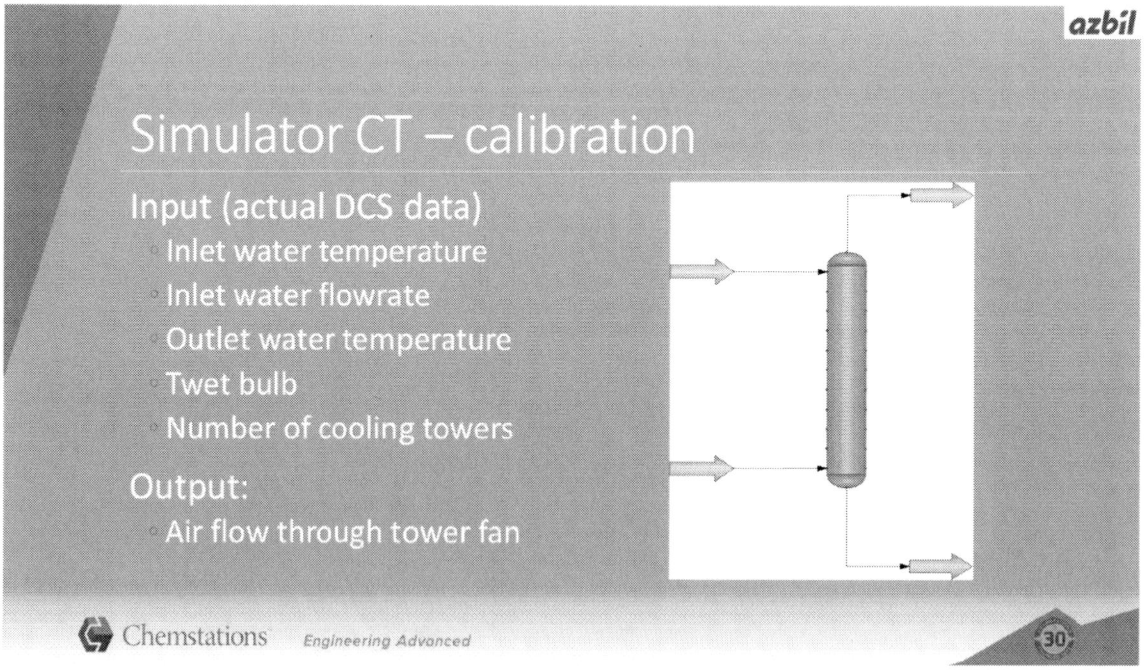
azbil
Simulator CT – calibration
Input (actual DCS data)
Inlet water temperature
Inlet water flowrate
Outlet water temperature
Twet bulb
Number of cooling towers
Output:
Air flow through tower fan
Chemstations
Engineering Advanced
30

azbil

Simulator CT - operation

Input

- Inlet water temperature
- Inlet water flowrate
- Twet bulb
- Number of cooling towers
- Air flow through tower fan

Output:

- Outlet water temperature

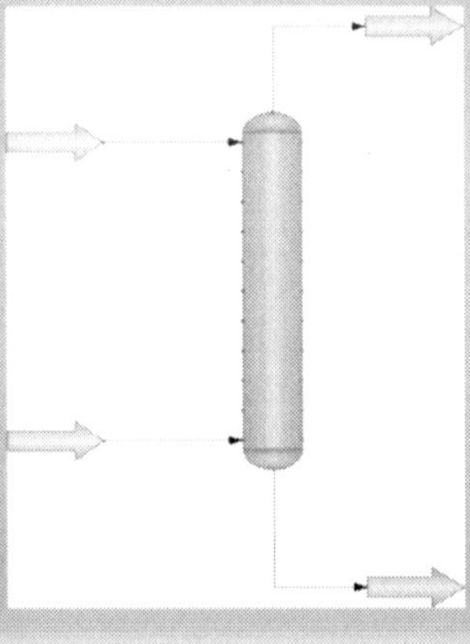

Chemstations Engineering Advanced

azbil

Validation of Cooling Tower model

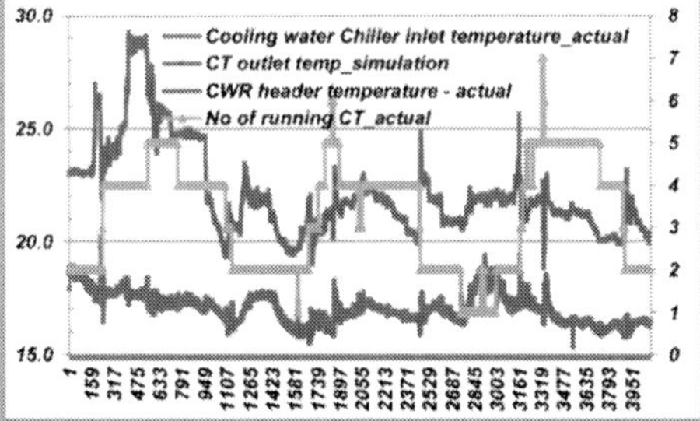

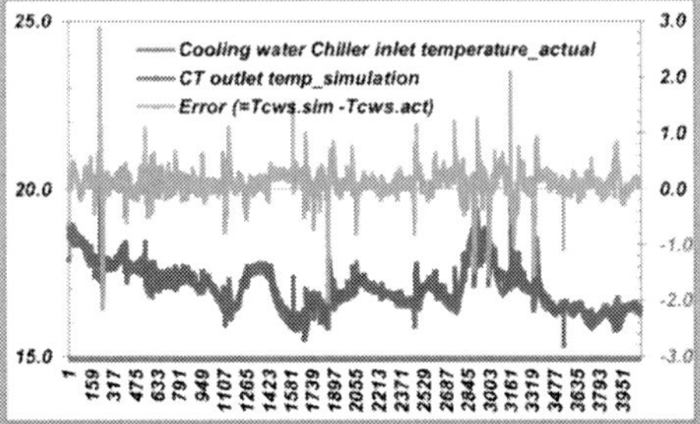

Chemstations Engineering Advanced

Chiller model (Regression Model)

azbil

(1) Chiller performance model:

$$\therefore COP_{chiller} = f(T_{cw,in}, F_{cw}, Q_{chw}, T_{chw,out})$$

$\therefore Q_{comp} = COP_{chiller} \times Q_{chw}$

Where, $COP_{chiller}$; Chiller performance index [kW/(GJ/h-chw)]

Q_{comp} ; Required compressor power [kW]

$T_{cw,in}$; Cooling water temperature entering condenser [degC]

F_{cw} ; Cooling water flow rate entering condenser [m3/h]

Q_{chw} ; Chilled water duty [GJ/h]

$T_{chw,out}$; Chilled water temperature leaving evaporator [degC]

<Pertubation model>:

$$\therefore \Delta COP_{chiller} = k_1 \times \Delta T_{cw,in} + k_2 \times \Delta F_{cw} + k_3 \times \Delta Q_{chw}$$

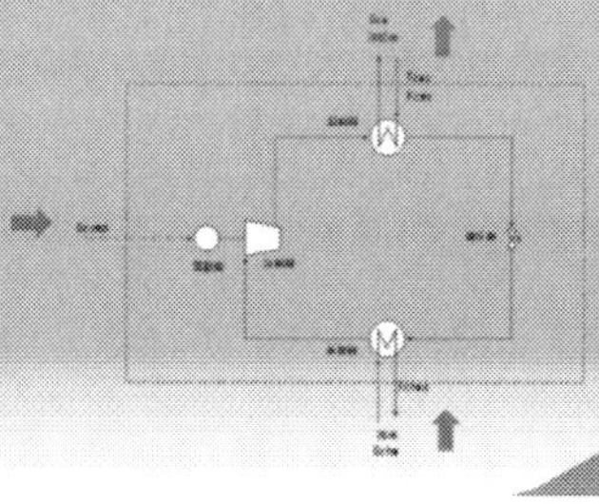

Chemstations Engineering Advanced

30

Chiller model – regression based

azbil

$$\therefore COP_{chiller} = f(T_{cw,in}, F_{cw}, Q_{chw}, T_{chw,out})$$

$\therefore Q_{comp} = COP_{chiller} \times Q_{chw}$

Where, $COP_{chiller}$; Chiller performance index [kW/(GJ/h-chw)]

Q_{comp} ; Required compressor power [kW]

$T_{cw,in}$; Cooling water temperature entering condenser [degC]

F_{cw} ; Cooling water flow rate entering condenser [m3/h]

Q_{chw} ; Chilled water duty [GJ/h]

$T_{chw,out}$; Chilled water temperature leaving evaporator [degC]

<Pertubation model>:

$$\therefore \Delta COP_{chiller} = k_1 \times \Delta T_{cw,in} + k_2 \times \Delta F_{cw} + k_3 \times \Delta Q_{chw}$$

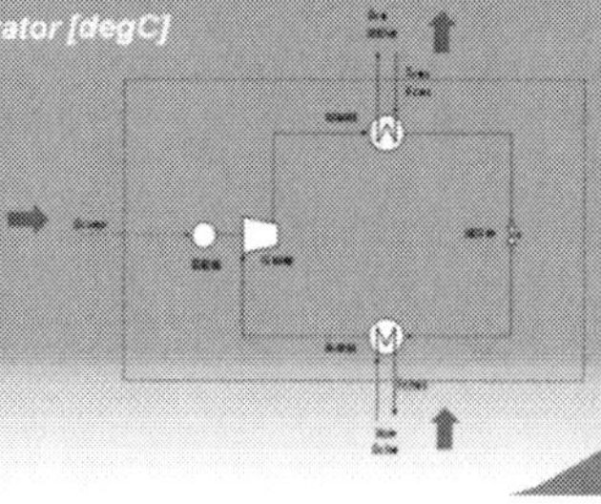

Chemstations Engineering Advanced

30

azbil

Hybrid – reduced and process simulator

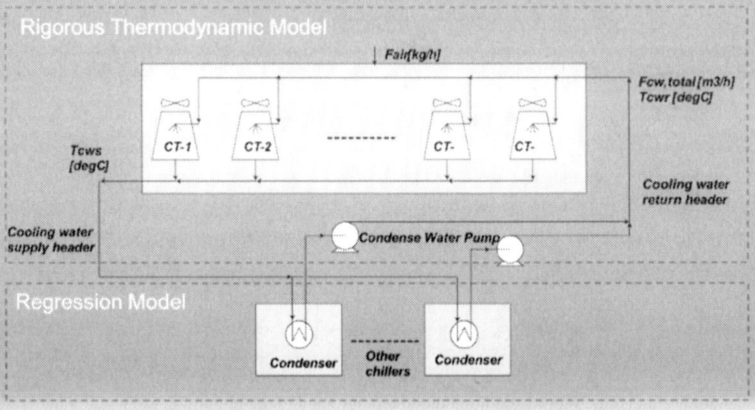

Chemstations Engineering Advanced

azbil

LP Optimizer

Take the equations and find the optimum

Chemstations Engineering Advanced

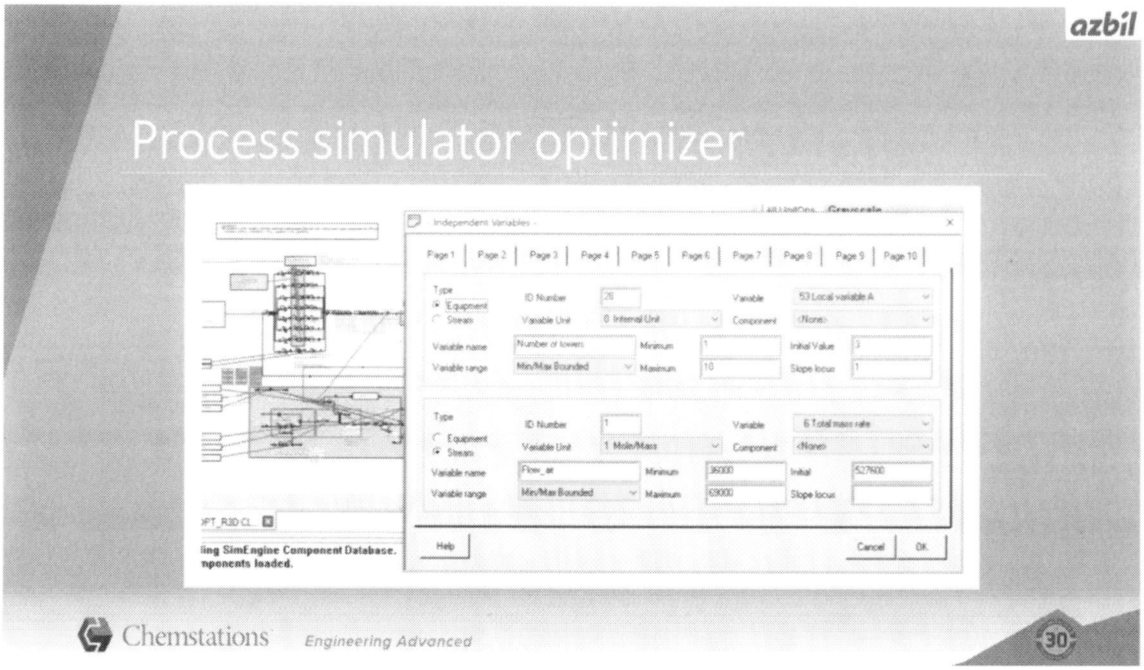
azbil
Process simulator optimizer
Independent Variables
Page 1
Page 2
Page 3
Page 4
Page 5
Page 6
Page 7
Page 8
Page 9
Page 10
Type
Equipment
Stream
ID Number
Variable
53 Local variable A
Variable Unit
0 Internal Unit
Component
<None>
Variable name
Number of towers
Minimum
Initial Value
Variable range
Min/Max Bounded
Maximum
Slope locus
6 Total mass rate
1 Mole/Mass
Flow_air
36000
Initial
527600
69000
Help
Cancel
OK
ling SimEngine Component Database.
mponents loaded.
Chemstations
Engineering Advanced
30

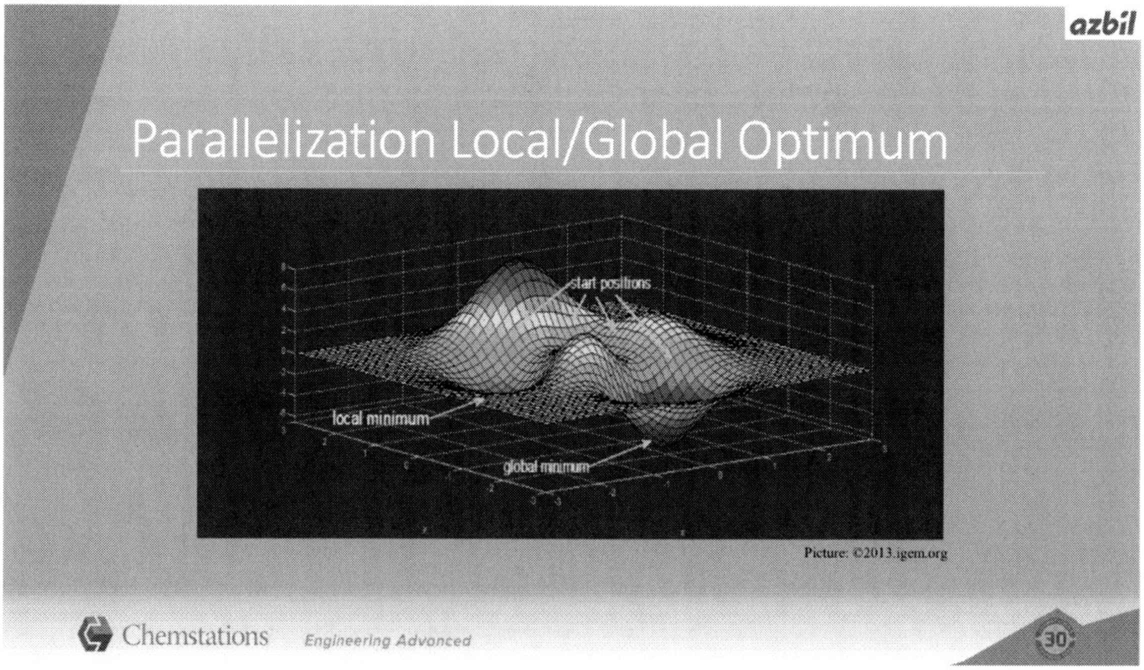
azbil
Parallelization Local/Global Optimum
start positions
local minimum
global minimum
Picture: ©2013.igem.org
Chemstations
Engineering Advanced
30

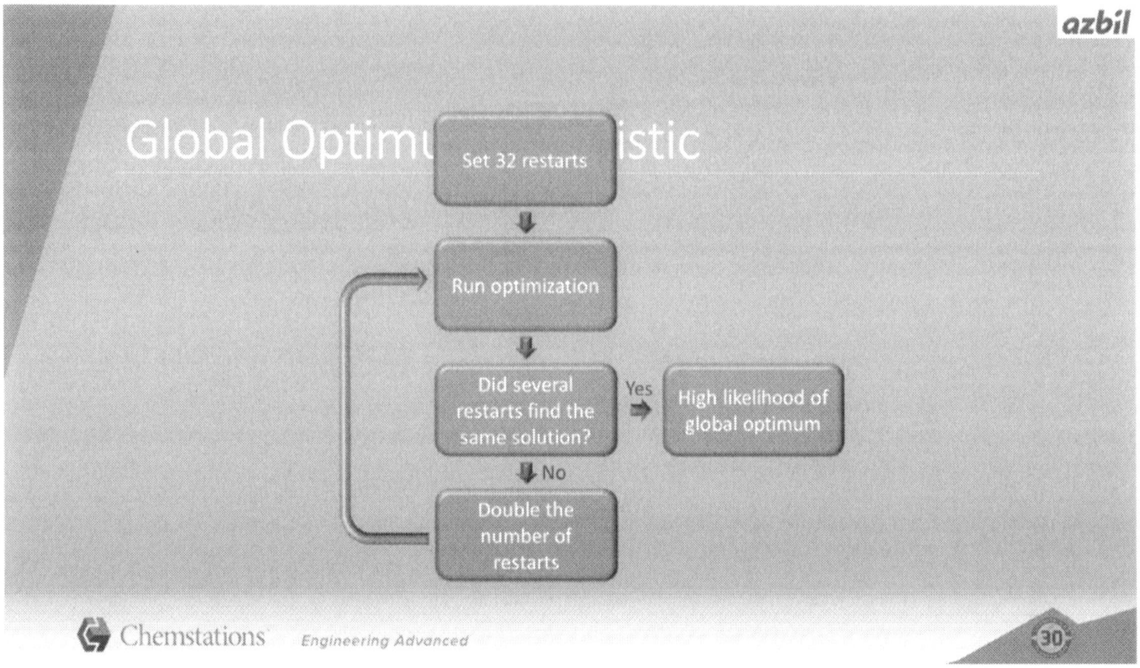
azbil
Global Optimu
istic
Set 32 restarts
Run optimization
Did several restarts find the same solution?
Yes
High likelihood of global optimum
No
Double the number of restarts
Chemstations
Engineering Advanced
30

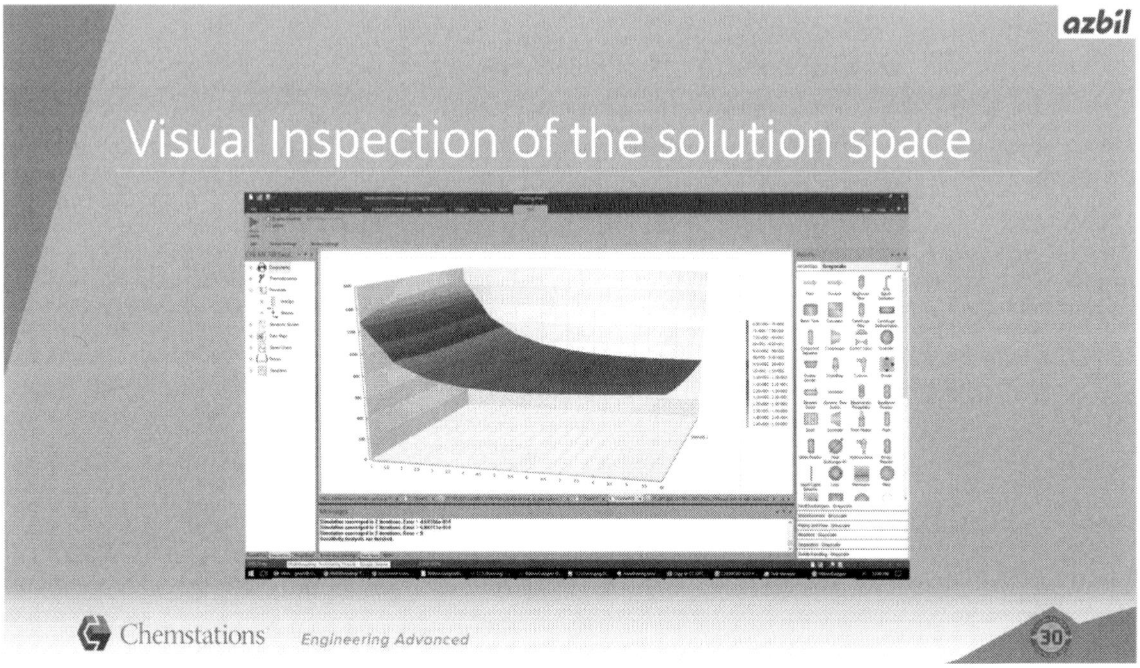
azbil
Visual Inspection of the solution space
Chemstations
Engineering Advanced
30

azbil

Online model – data connection

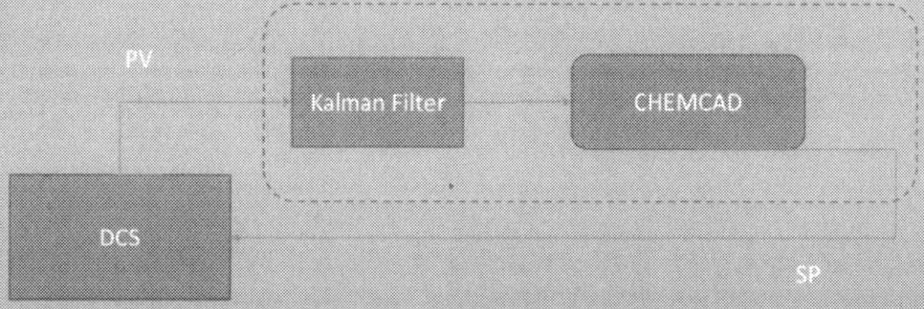

Chemstations Engineering Advanced

30

azbil

Results of optimization – energy savings

(1) Trend of System COP:

- System COP ratio is based on compressor, condenser pump, cooling tower fan powers. (System COP ratio = 1 is baseline operation. Smaller number is better)

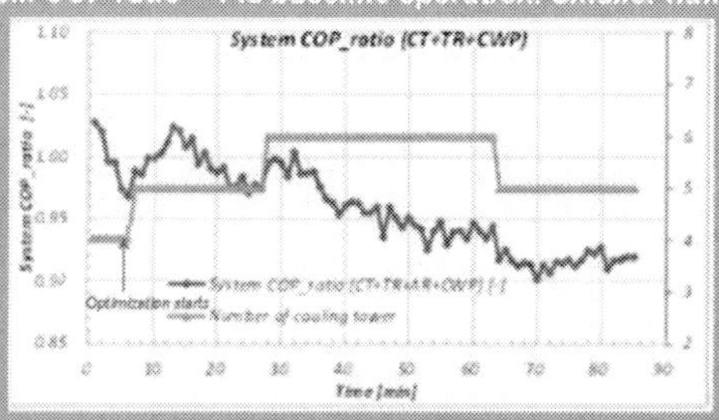

Chemstations Engineering Advanced

30

azbil

Results of Optimization – explanation of plot

- COP is a way to measure "energy cost to produce a unit of chilled water. For our definition of COP, a smaller COP is better
- The plot (previous slide) shows operations data and overall COP
  - At five minutes, the optimizer was allowed to optimized number of cooling towers
  - Approximately 50 minutes, the optimizer was allowed to optimize cooling water total flowrate
  - The optimizer was turned on one variable at a time to see the overall plot trend and convince the customer that both functions reduced COP
- An audience member asked about the cost of makeup water. This customer wasn't concerned with the makeup water cost. It would be simple to add a 'penalty function' to force the optimizer to be more conservative about makeup water

Chemstations Engineering Advanced

30

azbil

Sensitivity Study – effect of n cooling tower on System COP1. Introduction

(1) Process conditions:

- ***Chilled water demand: = 17.09[GJ/h] x 75[%] x 2 [chillers] (constant)***
- ***Number of cooling tower = 3 to 7 (Base case : Nct =4)***
- ***Cooling water flow rate entering condenser = 500 [m3/h] (constant)***
- ***Twet = 14.0, 15.5, 16.9 [degC] (Base case : Twet = 15.5.[degC])***

(2) Calculation Result:

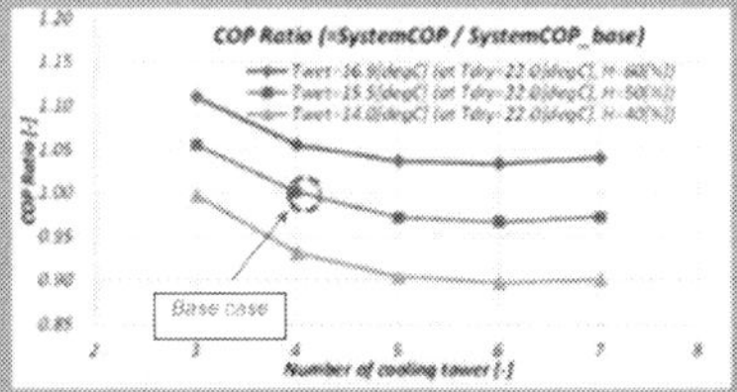

- ***Chance to improve system COP by increasing the number of cooling tower.***
- ***Wet-bulb temperature change of 1.5 [degC] gives system COP change of 5[%].***

Chemstations Engineering Advanced

30

azbil

The idea of a hybrid model

- Graphic combining
  - Rigorous column model
  - Regressed chiller and pump
  - Perturbation optimizer capable – mixed integer capable

Chemstations *Engineering Advanced*

azbil

Setup Time Comparison

Chemstations *Engineering Advanced*

azbil

Comparison of LP vs Simulator optimization

	LP Optimization	Simulator Optimization
Calculation speed	Fast	Bring a powerful computer
Reliability	Reduced models aren't accurate	Virtual model of the plant
Build	Requires expert use of equations and programming and lots of time	Fast. Requires skills taught to undergrads
Initial cost	High. Step change of facility, site visits, custom programming	Software and modeling
Lifecycle cost	Depends on maintenance	Annual cost of software, plus maintenance of tool

Chemstations Engineering Advanced 30

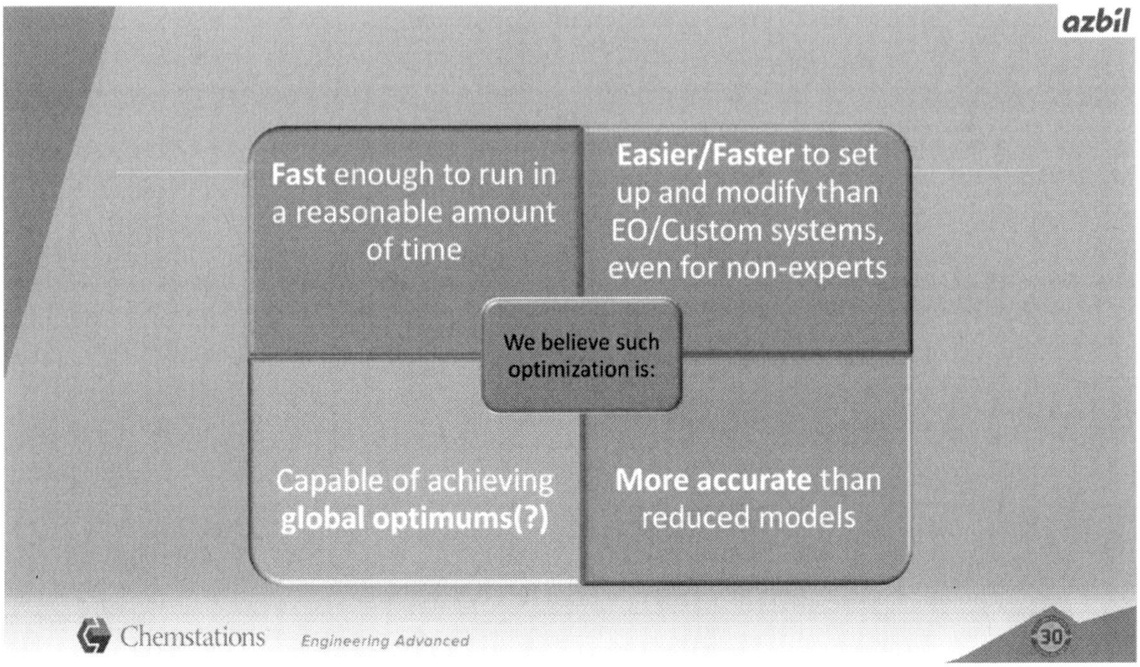

azbil

Summary and Q&A

Process simulator model combines rigorous unit operations with reduced models

Online optimization without analytical derivatives

Faster time-to-build expands application of online optimization

Case result: online optimization of chilled water network performs well, was easy to build, is more accurate for cooling tower

Chemstations *Engineering Advanced*

azbil

Questions from the audience

Q: Does the model consider the cost of makeup water? We have a facility in the desert

A: We used an objective function which totaled the energy from different sources on the flowsheet. We could easily change the objective function to consider cost of energy and cost of our water. Our model did not consider the cost of makeup water because that wasn't a concern for this customer.

Q: Which method is better for optimization? Traditional LP of reduced equations or a simulator based perturbation?

A: Our opinion is that a large number of engineers are capable of building a process simulation and then using the simulator's optimizer tool; only a small number of engineers are capable of building a reduced model and an LP optimizer. This resource constraint impacts the cost of building optimizations. See our comparison of LP vs simulator optimization slide for more thoughts.

Chemstations *Engineering Advanced*

Want to know more?

Contact Mike Suzuki of Azbil for information about projects like this.

Visit for information about CHEMCAD.

azbil

Chemstations Engineering Advanced

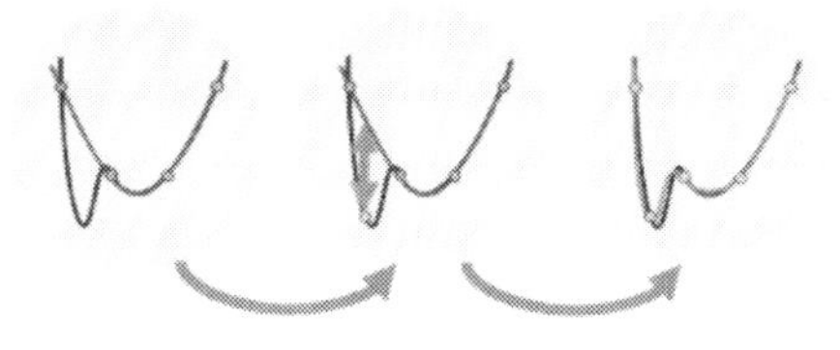

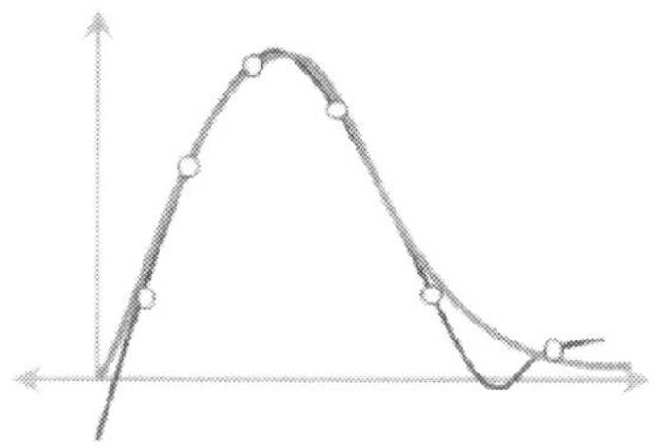

ALAMO: Machine learning from data and first principles

Zachary Wilson, Nick Sahinidis
Carnegie Mellon University

PROCESS DEVELOPMENT SYMPOSIUM 2018

2018 Process Development Symposium | June 5-7, 2018
Hyatt Lodge at Hamburger University | Oak Brook, IL

Process Network Optimization

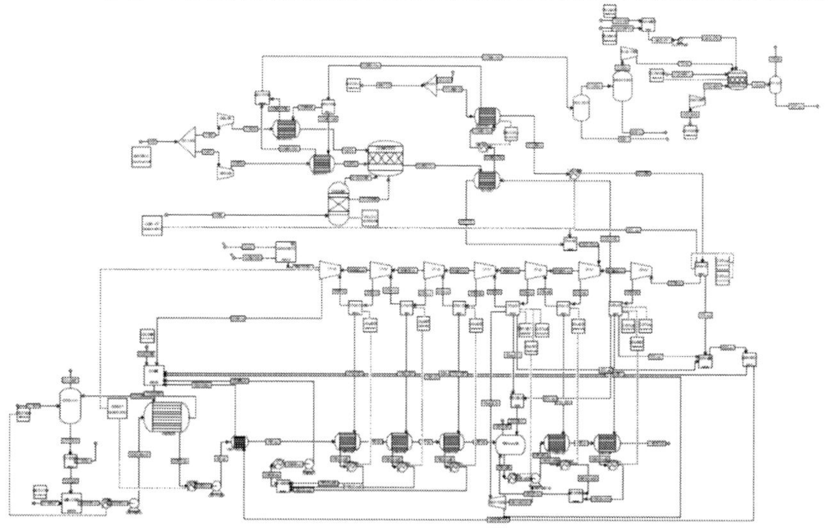

Pulverized coal plant Aspen Plus® simulation provided by the National Energy Technology Laboratory

PROCESS DEVELOPMENT SYMPOSIUM 2018

2018 Process Development Symposium | June 5-7, 2018
Hyatt Lodge at Hamburger University | Oak Brook, IL

Challenges

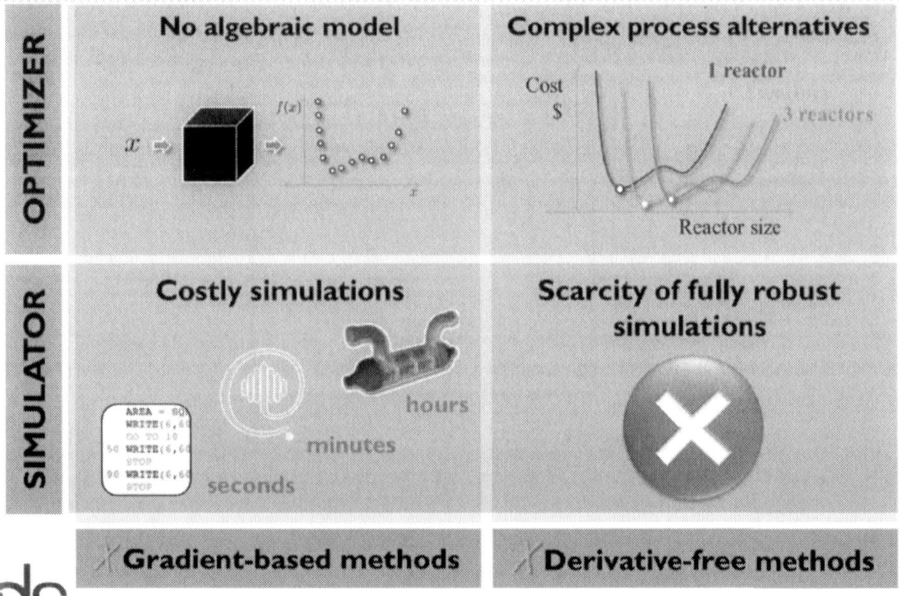

Problem Decomposition

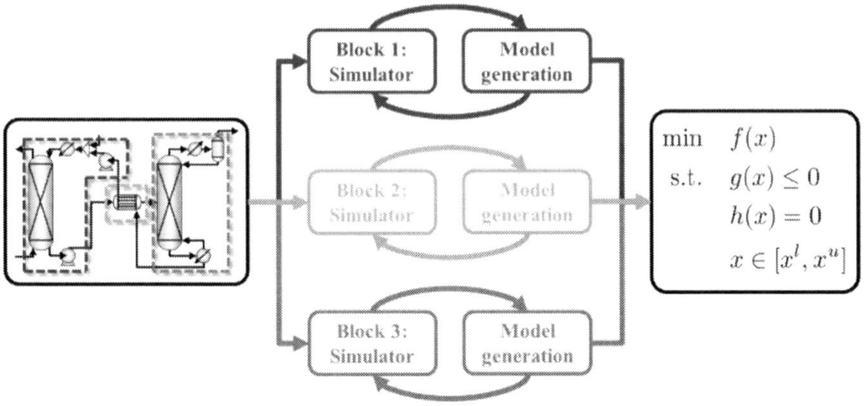

Process Simulation

Decompose network into process blocks

Surrogate Models

Build simple and accurate models with a functional form tailored for an optimization framework

Optimization Model

Add algebraic constraints design specs, heat/mass balances, and logic constraints

Machine Learning Problem

Build a model of output variables $\mathbf{z}$ as a function of input variables x over a specified interval

$$x \in \mathbb{R}^k \qquad x^l \leq x \leq x^u \qquad \begin{pmatrix} x_1 \\ x_2 \\ \vdots \\ x_j \\ \vdots \\ x_k \end{pmatrix}$$

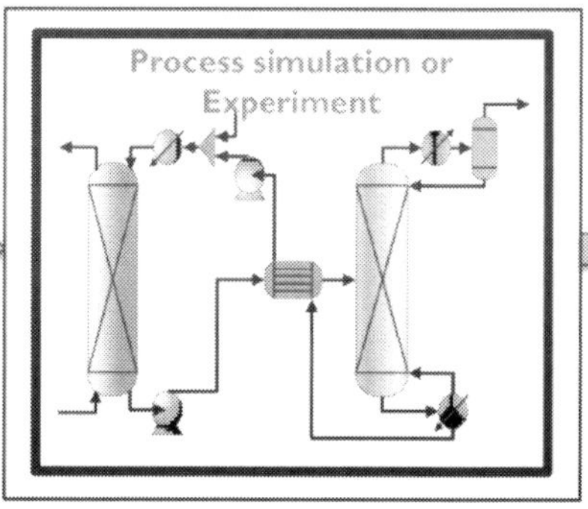

$$\begin{pmatrix} z_1 \\ z_2 \\ \vdots \\ z_l \\ \vdots \\ z_m \end{pmatrix} \qquad z \in \mathbb{R}^m \qquad z = f(x)$$

Independent variables:
Operating conditions, inlet flow properties, unit geometry

Dependent variables:
Efficiency, outlet flow conditions, conversions, heat flow, etc.

Desired Model Attributes

1. **Accurate**
   - We want to reflect the true nature of the system
2. **Simple**
   - Usable for algebraic optimization
   - Interpretable
3. **Generated from a minimal data set**
   - Reduce experimental and simulation requirements
4. **Obeys physics and user insights**
   - Increase fidelity and validity in regions with no measurements

ALAMO

Automated Learning of Algebraic MOdels

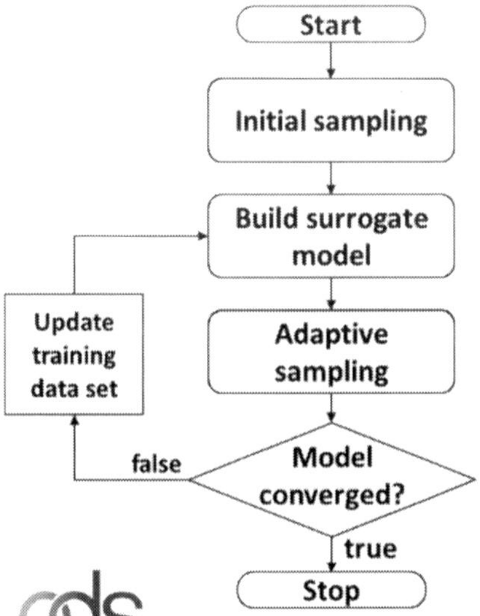

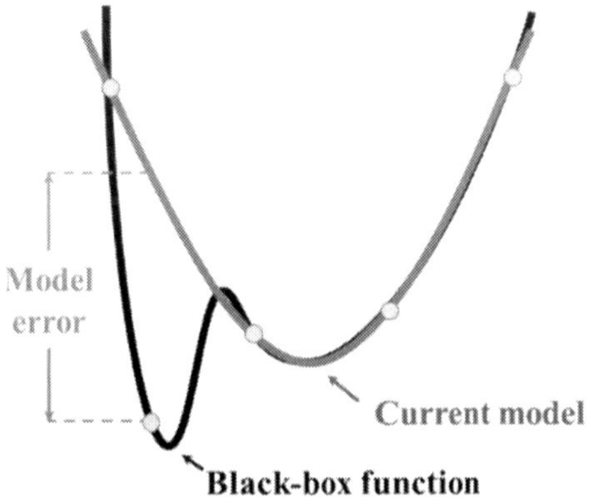

2018 Process Development Symposium | June 5-7, 2018
Hyatt Lodge at Hamburger University | Oak Brook, IL

Overfitting and True Error

- Step 1: Define a large set of potential basis functions

$$\hat{z}(x) = \beta_0 + \beta_1 x_1 + \beta_2 x_2 + \beta_3 x_1 x_2 + \beta_4 e^{x_1} + \beta_5 e^{x_2} + \ldots$$

- Step 2: Model reduction

$$\hat{z}(x) = 2 + x_2 + 5\,e^{x_1}$$

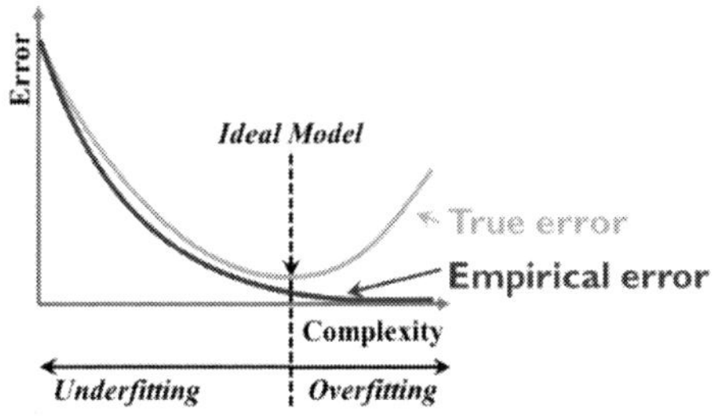

2018 Process Development Symposium | June 5-7, 2018
Hyatt Lodge at Hamburger University | Oak Brook, IL

ALAMO Provides Simple Models

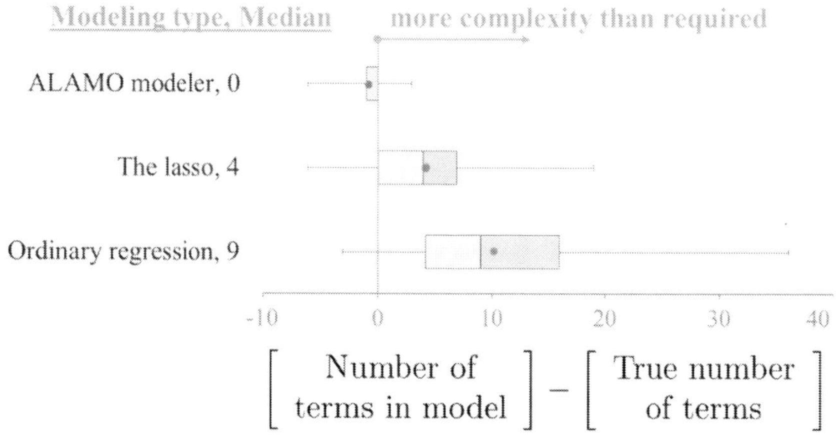

Results over a test set of 45 known functions treated as black boxes with bases that are available to all modeling methods

Error Maximization Sampling

- Search the problem space for areas of model inconsistency or model mismatch
- Find points that maximize the model error with respect to the independent variables

Surrogate model

$$\max_{x} \left(\frac{z(x) - \hat{z}(x)}{z(x)} \right)^2$$

  - Optimized using derivative-free solver SNOBFIT (Huyer and Neumaier, 2008)
  - SNOBFIT outperforms most derivative-free solvers (Rios and Sahinidis, 2013)

Constrained Regression

$$0 \leq [\mathrm{A}]_t \leq [\mathrm{A}]^{\max}$$

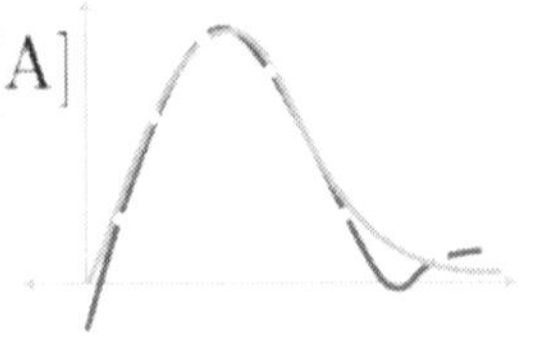

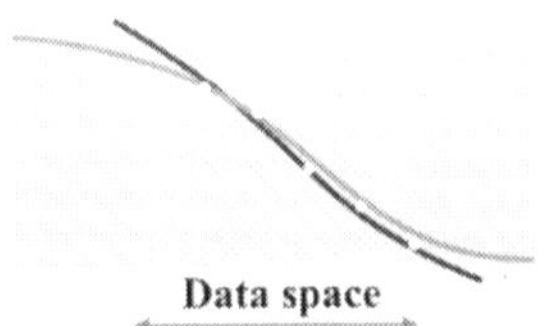

Constrained regression relies on global optimization technology (BARON)

Constrained Regression Formulation

Standard regression

$$\min_{\beta_1,\beta_2} \sum_{i=1}^{4} (z_i - \hat{z}(x_i;\beta_1,\beta_2))^2$$

Surrogate model

Constrained regression

easy:

$$\min_{\beta_1,\beta_2} \sum_{i=1}^{4} (z_i - \hat{z}(x_i;\beta_1,\beta_2))^2$$
$$\text{s.t.} \quad \beta_1 \geq \beta_2$$

tough:

$$\min_{\beta_1,\beta_2} \sum_{i=1}^{4} (z_i - \hat{z}(x_i;\beta_1,\beta_2))^2$$
$$\text{s.t.} \quad \hat{z}(x_i;\beta_1,\beta_2) \geq 0 \quad \forall x$$

- Challenging due to the semi-infinite nature of the regression constraints
- Use intuitive restrictions among predictor and response variables to infer nonintuitive relationships between regression parameters

2018 Process Development Symposium | June 5-7, 2018
Hyatt Lodge at Hamburger University | Oak Brook, IL

Implied Parameter Restrictions

Find a model $\hat{z}$ such that $\hat{z}(x) \geq 0$ with a fixed model form:

$$\hat{z}(x) = \beta_1 x + \beta_2 x^3$$

Step 1: Formulate constraint in z- and x-space

$$\min_{\beta_1, \beta_2} \sum_{i=1}^{4} \left(z_i - [\beta_1 x + \beta_2 x^3]\right)^2$$

$$\text{s.t.} \quad \beta_1 x + \beta_2 x^3 \geq 0 \quad x \in [0, 1]$$

1 parametric constraint

Step 2: Identify a sufficient set of β-space constraints

$$\min_{\beta_1, \beta_2} \sum_{i=1}^{4} \left(z_i - [\beta_1 x + \beta_2 x^3]\right)^2$$

$$\text{s.t.} \quad \begin{cases} 0.240\, \beta_1 + 0.0138\, \beta_2 \geq 0 \\ 0.281\, \beta_1 + 0.0223\, \beta_2 \geq 0 \\ 0.120\, \beta_1 + 0.00173\, \beta_2 \geq 0 \\ 0.138\, \beta_1 + 0.00263\, \beta_2 \geq 0 \end{cases}$$

4 β-constraints

Global optimization problems solved with BARON

2018 Process Development Symposium | June 5-7, 2018
Hyatt Lodge at Hamburger University | Oak Brook, IL

Types of Restrictions

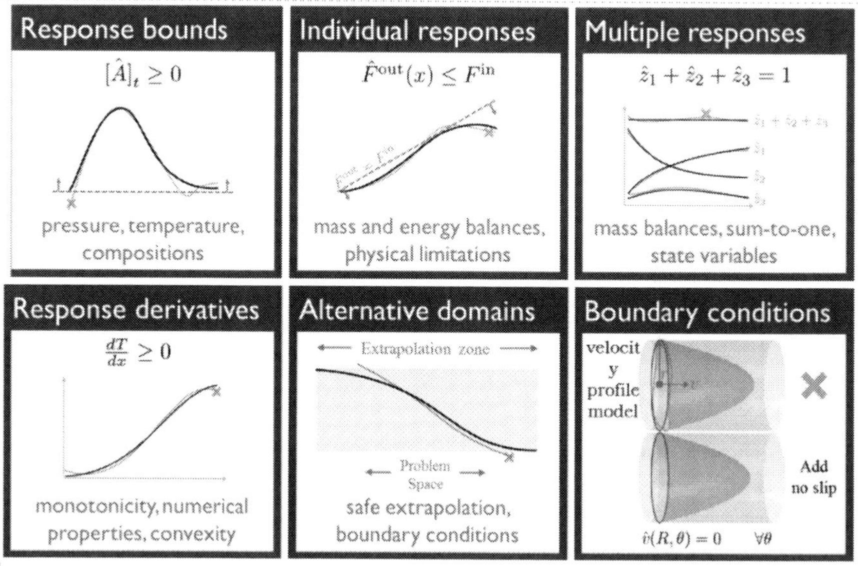

2018 Process Development Symposium | June 5-7, 2018
Hyatt Lodge at Hamburger University | Oak Brook, IL

Carbon Capture System Design

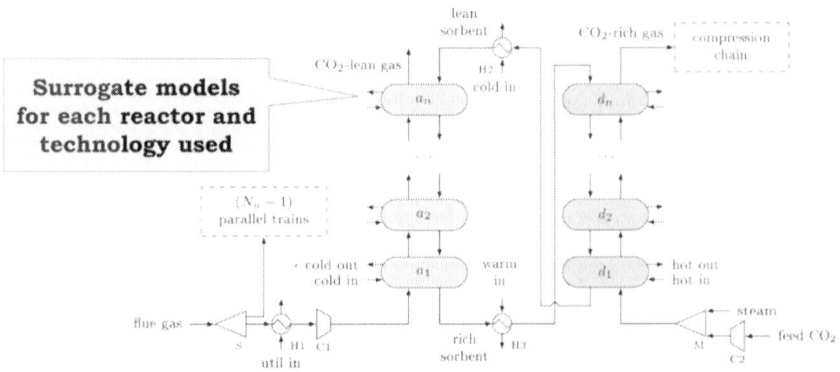

- Discrete decisions: How many units? Parallel trains? What technology used for each reactor?
- Continuous decisions: Unit geometries
- Operating conditions: Vessel temperature and pressure, flow rates, compositions

Bubbling Fluidized Bed Adsorber

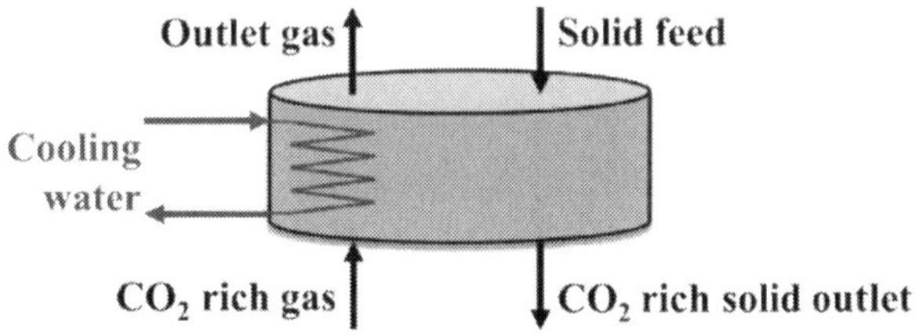

- Model inputs (16 total)
  - Geometry (3)
  - Operating conditions (5)
  - Gas mole fractions (2)
  - Solid compositions (2)
  - Flow rates (4)
- **Model outputs (14 total)**
  - **Geometry required (2)**
  - **Operating condition required (1)**
  - **Gas mole fractions (3)**
  - **Solid compositions (3)**
  - **Flow rates (2)**
  - **Outlet temperatures (3)**

Model created by Andrew Lee at the National Energy Technology Laboratory

Bubbling Fluidized Bed: Case study

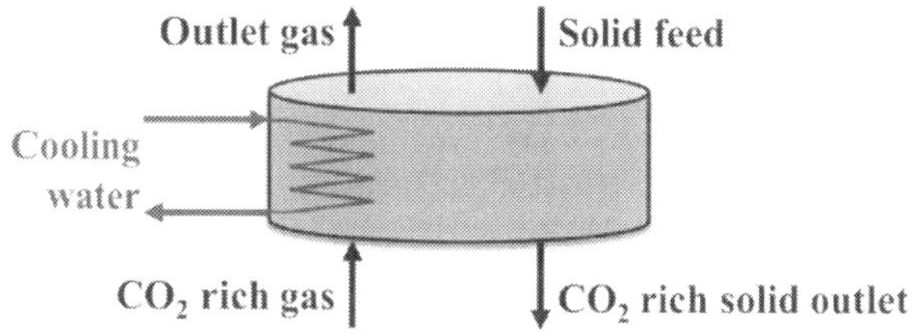

Goal: Optimize a bubbling fluidized bed reactor by

- **Minimizing the increased cost of electricity**
- **Maximizing CO_2 removal**

Bubbling Fluidized Bed: Case study

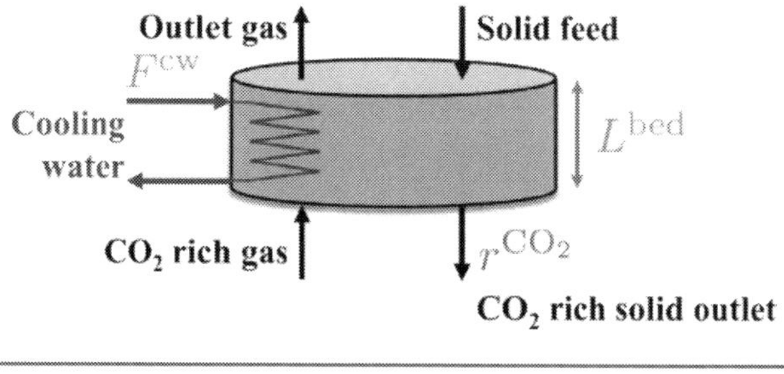

Generate model: $r^{\mathrm{CO_2}}(L^{\mathrm{bed}}, F^{\mathrm{cw}}) = f_1(L^{\mathrm{bed}}, F^{\mathrm{cw}})$

Over the Range:

$$\begin{aligned} 1\ \mathrm{m} &\le L^{\mathrm{bed}} \le 10\ \mathrm{m} \\ 1 \cdot 10^4\ \tfrac{\mathrm{kmol}}{\mathrm{h}} &\le F^{\mathrm{cw}} \le 20 \cdot 10^4\ \tfrac{\mathrm{kmol}}{\mathrm{h}} \end{aligned}$$

Potential Model

$$
\begin{aligned}
r^{\mathrm{CO_2}} = {} & \beta_0 + \beta_1 F + \beta_2 L + \beta_3 \mathrm{e}^F + \beta_4 \mathrm{e}^L + \beta_5 \ln(F) + \beta_6 \ln(L) + \beta_7 \tfrac{F}{L} + \beta_8 \tfrac{2\,F}{L^2} + \\
& \beta_9 F\,L^2 + \beta_{10} F^2\,L + \beta_{11} \tfrac{F}{\sqrt{L}} + \beta_{12} F\,\sqrt{L} + \beta_{13} \tfrac{L}{\sqrt{F}} + \beta_{14} \sqrt{F}\,L + \beta_{15} \tfrac{F}{L^{\frac{1}{3}}} + \beta_{16} F\,L^{\frac{1}{3}} + \\
& \beta_{17} \tfrac{L}{F^{\frac{1}{3}}} + \beta_{18} F^{\frac{1}{3}}\,L + \beta_{19} 2\,F\,\ln(F) + \beta_{20} F\,\ln(L) + \beta_{21} L\,\ln(F) + \beta_{22} L\,\ln(L) + \beta_{23} \tfrac{1}{F} + \\
& \beta_{24} \tfrac{1}{F^2} + \beta_{25} F^2 + \beta_{26} \tfrac{1}{F^3} + \beta_{27} \tfrac{1}{\sqrt{F}} + \beta_{28} \sqrt{F} + \beta_{29} F^3 + \beta_{30} \tfrac{1}{F^{\frac{1}{3}}} + \beta_{31} F^{\frac{1}{3}} + \beta_{32} \tfrac{1}{L} + \\
& \beta_{33} \tfrac{1}{L^2} + \beta_{34} L^2 + \beta_{35} \tfrac{1}{L^3} + \beta_{36} \tfrac{1}{\sqrt{L}} + \beta_{37} \sqrt{L} + \beta_{38} L^3 + \beta_{39} \tfrac{1}{L^{\frac{1}{3}}} + \beta_{40} L^{\frac{1}{3}} + \beta_{41} \tfrac{1}{F\,L} + \\
& \beta_{42} \tfrac{2\,L^2}{F} + \beta_{43} \tfrac{\sqrt{L}}{F} + \beta_{44} \tfrac{F^2}{L^2} + \beta_{45} F^2\,L^2 + \beta_{46} \tfrac{\sqrt{F}}{L} + \beta_{47} \tfrac{F^3}{L} + \beta_{48} \tfrac{1}{F^3\,L^3} + \beta_{49} \tfrac{1}{\sqrt{F}\,\sqrt{L}} + \\
& \beta_{50} \tfrac{\sqrt{F}}{\sqrt{L}} + \beta_{51} \sqrt{F}\,\sqrt{L} + \beta_{52} \tfrac{F^3}{L^3} + \beta_{53} F^3\,L^3 + \beta_{54} \tfrac{L^{\frac{1}{3}}}{F} + \beta_{55} \tfrac{F^{\frac{1}{3}}}{L} + \beta_{56} \tfrac{1}{F^{\frac{1}{3}}\,L^{\frac{1}{3}}} + \beta_{57} \tfrac{F^{\frac{1}{3}}}{L^{\frac{1}{3}}} + \\
& \beta_{58} F^{\frac{1}{3}}\,L^{\frac{1}{3}} + \beta_{59} F^{\frac{1}{3}}\,\ln(L) + \beta_{60} L^{\frac{1}{3}}\,\ln(F) + \beta_{61} L^{\frac{1}{3}}\,\ln(L) + \beta_{62} F\,L
\end{aligned}
$$

66 basis functions

Not tractable in an algebraic superstructure formulation!

Bubbling Fluidized Bed: Results

$$r = 0.2L^{\frac{1}{3}} - 0.1LF^{-\frac{1}{2}} - 0.1L\log(F) + 0.05F^{\frac{1}{2}}\log(L)$$

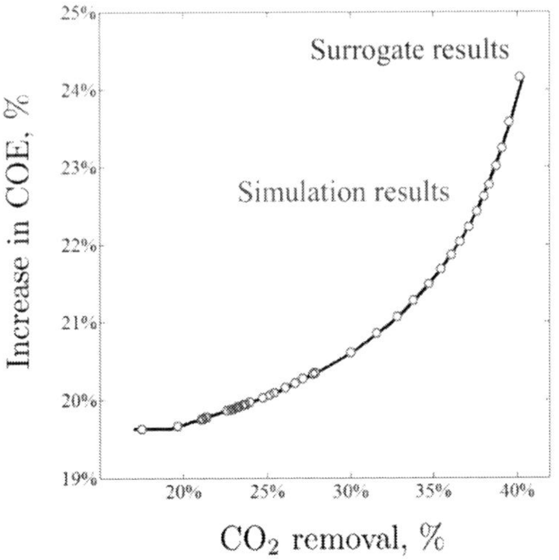

PROCESS DEVELOPMENT SYMPOSIUM 2018
2018 Process Development Symposium | June 5-7, 2018
Hyatt Lodge at Hamburger University | Oak Brook, IL

Conclusions

- ALAMO provides algebraic models that are
  - ✓ Accurate
  - ✓ Simple
  - ✓ Generated from a minimal number of data points
- ALAMO's constrained regression facility allows modeling of
  - ✓ Bounds on response variables
  - ✓ Logical relationships between variable groups
- Built on top of state-of-the-art optimization solvers
- Available (free) from www.minlp.com

Downstream Technology

bp

AIChE PDS Symposium - 2018

The Practice of Simulation in Process Engineering and Optimization

Kheng Lau
Simulation Advisor
Refining Technology and Engineering

Presentation Agenda

bp

- Overview of Simulation Software Progressions
- Application Examples
  - Steady State
  - Dynamics
- Closing Remarks

Downstream Technology 2

Process Simulation Software Packages

Sequential Modular (SM)

- Easy to use for beginners
- Robust convergence algorithms for different process equipment
- Different levels of rigor in modeling process equipment
- Large selection of rigorous physical property models capturing VLE/LLE behavior

Equation-Oriented (EO)

- Generally used by simulation specialists
- Fast convergence with flowsheets having large recycle flows and complex heat integration
- Widely used in closed-loop online optimization
- Flexibility in problem specification

Progression of Simulation Over the Years

SM Simulator

EO Simulator

Text Based Input File

GUI

- Rigor of equipment models
  - Rate-based column models
- Integrating equipment design and simulation
- Solver technologies
- Online use in operations
- Management of data I/O for simulation applications
- Industry standards (OPC)
- Human-Machine-Interface

Digital Technology

1980's 1990's 2000's 2010's Now...

VAX/IBM/HP Workstation

Desktops Laptops

Multi-core Technologies

Cloud Hosting

Tablets

- **Computing Hardware Shift**

One Key Aspect of Process Simulation Software Changes

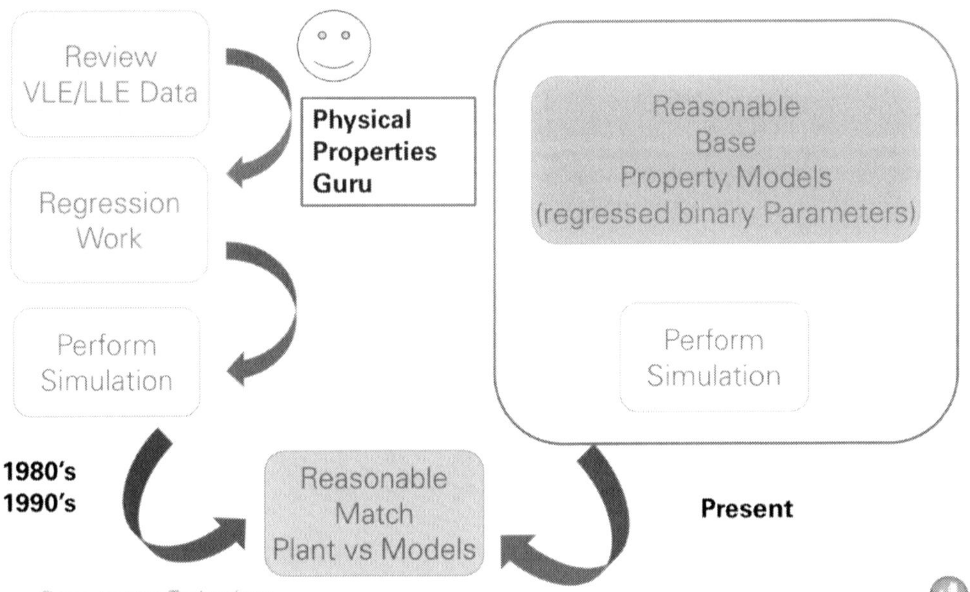

Downstream Technology

Process Engineering Applications

- Hydrate formation in exchangers
  - Can we avoid the conditions that hydrates form particularly in the winter?
    - What thermodynamic inhibitors can we use? How much?
- Non-ideal separation of mercaptans in LPG
- Evaluating performance of columns
  - Multi-draw fractionators
    - Crude/Vac/FCC/Coker/Hydrocracker
      - Equilibrium stages in different sections of columns
  - Tray and packing loading
- Revamp studies
  - Process changes like re-routing of H_2 or fuel gas streams in different parts of the refinery

Downstream Technology

Refinery LP Applications

- LP is used in representing a refinery for planning and scheduling
  - How do we update the LP?
    - Process models come into play in generating data for updating the LP
      - Accurate rigorous process models are critical
    - Process data generally do not have enough operating ranges for the LP
- Refinery configuration studies
  - How should I run my refinery if I add a new unit?
    - Do I need to revamp other units to take advantage of the new unit? How does this affect current crude diet?
  - Process models with rigorous reactor models are used to do such engineering analysis now instead of relying on LP's or other simplified representation of the refinery

Closed Loop Optimization in Operations

- Widely used in PetroChem and Refining primarily with EO technology
- Rigorous steady state process models are used
  - Columns/Reactors/Compressors/Exchangers/etc.
  - Capturing key constraints consistent with advanced controllers
- Process and lab data are retrieved automatically and sent to process models
  - Averaging and bad data detection
  - Steady state detection
  - Process model is solved twice during an optimization cycle
    - Matching process data
    - Optimization
- Results from optimization are sent to DCS as targets for advanced controllers to move the plant

Plant DCS Architecture for Online Applications

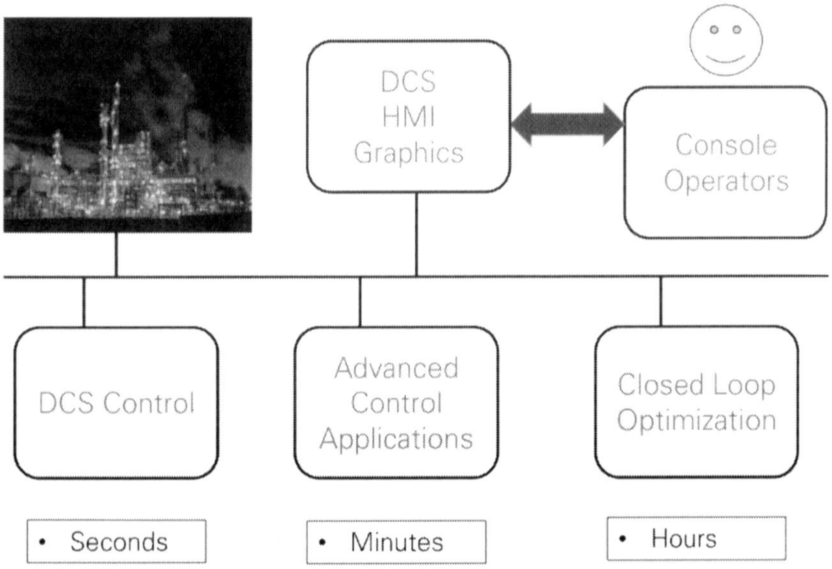

Downstream Technology

Olefins Plant Closed-Loop Optimizers

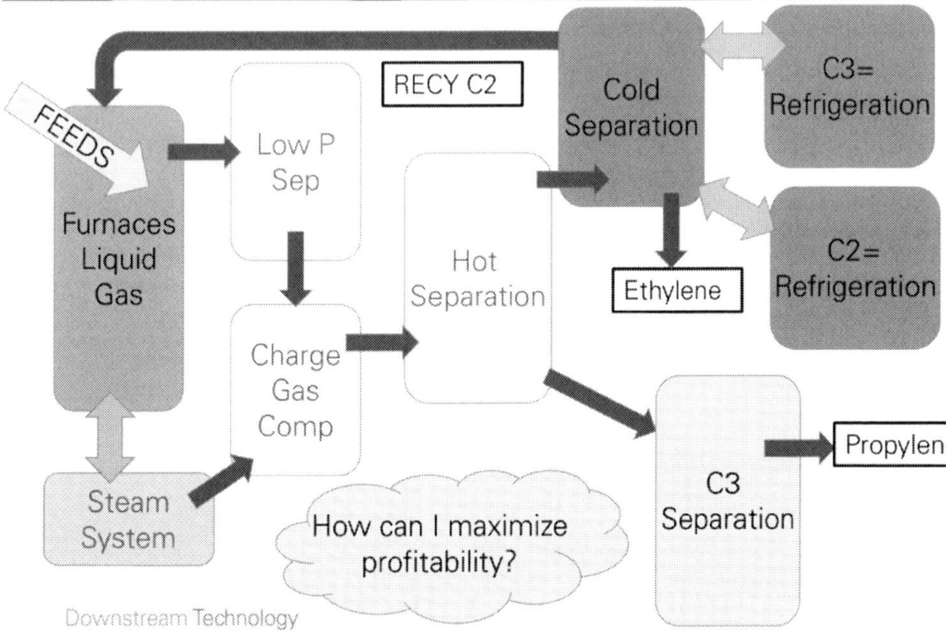

Downstream Technology

Factors Driving Ease of Deploying Online Applications

- Standard communication protocols
  - Ethernet
  - OLE for Process Control (OPC)
    - No need for custom interface development
      - Initial cost and future maintenance
        - Could take 2-3 months to develop a new interface
- Maturing of EO technology
  - Robust solvers and improvement in diagnostics
    - How do you convey the message of a singular matrix?
      - Actionable messages and better tools
  - Ease of building rigorous steady state process models
    - Use of GUI
- Lower cost of computing hardware

Downstream Technology

Dynamic Simulation Applications

- Representation of process units (Coker/etc) during a flaring event
  - Equipment
    - Columns/pumps/compressors/valves/reactors/exchangers/etc.
  - Relief loads and valve capacities
  - Evaluate strategies in routing light gases
- Training simulators
  - Rigorous representation of process units and automation
    - Compressor Controls/SIS/PLC/etc.
    - Equipment data and kinetic models for reactors
  - Performance
    - Calculation is typically 1.5 - 2 times faster than real time
  - Component list is generally less than the one used for steady state simulation even today (Refining)

Downstream Technology

bp

Components of High Fidelity Simulators

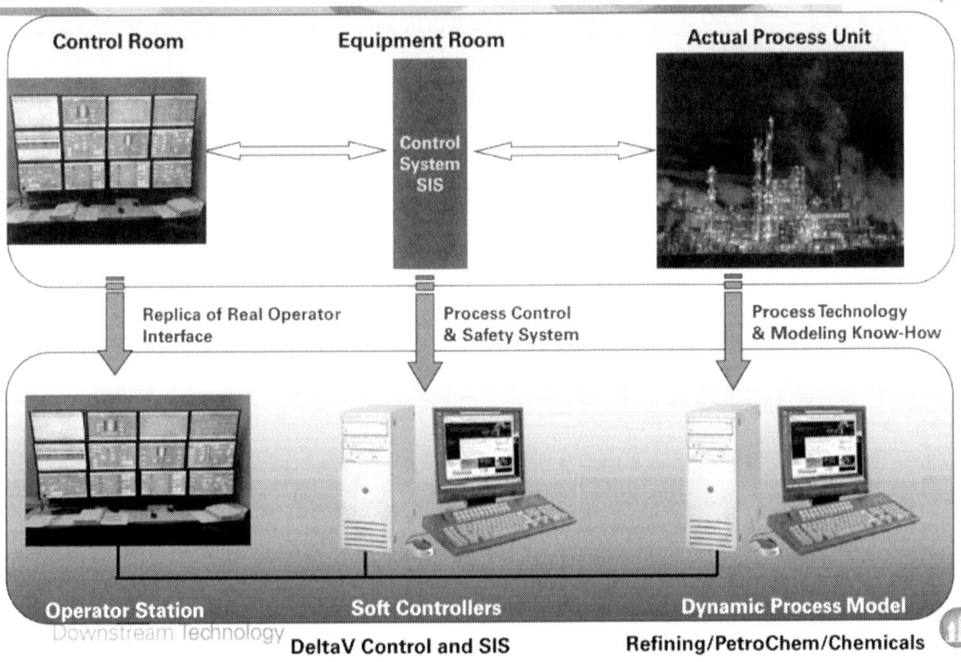

Downstream Technology

bp

Use of Training Simulators

- Training board operators
  - Unit operating procedures
    - Active learning environment
      - Instead of being told what would happen if one opens a by-pass valve too soon or late
      - Practice managing emergency events
  - Improving performance of board operators
    - Refresher training before start-up or shutdown
    - Preparation for emergency events
      - Hard to get in a live unit
- HMI and automation
  - Full test of DCS graphics executing procedures
  - SIS logics and resets including navigation on the DCS graphics

Downstream Technology

Key Benefits - Adoption of Windows based Computing Platform

- Cost dropped significantly while gaining computing power
- Typical closed loop online optimization applications changed from taking hours (1-1.5) to minutes (15-30) from start to finish
  - Solving two problems (100K-300K equations)
    - Matching process performance
    - Optimization
  - Readily available desktops or laptops are capable of such simulation
- Increase of rigor in high fidelity training simulators while keeping performance
  - Rigorous modeling of process equipment
    - Tray-by-Tray/Performance curves of compressors and pumps/etc.
  - Eliminate "It is just a simulator" mentality
    - Dynamic representation of the process unit that one can use to execute startup/shutdown/emergency procedures or develop new procedures

Downstream Technology

Closing Remarks

- Process simulation is a representation of our understanding of the process
  - Being able to model a process accurately is a testament of our understanding of the process
  - Without a good process model, it limits our capabilities in doing process simulation work
    - Innovation in process design
    - New ideas or better ways on how to run process units
    - Troubleshooting
    - Engineering analysis
  - Creating process unit "digital twin"
- Developing expertise in process simulation is a challenge for many operating companies
  - Combination of process understanding/control/simulation

Downstream Technology

16

Application of Computational Fluid Mechanics (CFD) and Discrete Element Methods (DEM) in Pharmaceutical Process Development

Authors:

Sinha, K., AbbVie Inc.

Sequential modular and equation-based simulation technologies are used widely nowadays. Both of these technologies play significant roles in steady state and dynamic simulation. Process simulation software has evolved in the last 20+ years. It is certainly easier to develop for example a distillation column model that matches process data fairly closely today. Advances in computing hardware and industrial standards have certainly transformed how simulation work is done. Process simulation enables many engineers around the world analyze and solve many engineering problems that they encounter. In addition, engineers use process simulation to evaluate new ideas in improving a process design or running a process unit. The process model used in simulation work must meet requirements defined for the intended application. Steady state and dynamic simulation application examples will be shared to illustrate important advances in simulation technology that make these applications possible.

FCC Regenerator Design and Improvement Using CFD and Kinetic Modeling Tools

Authors:

Davydov, L., UOP/Honeywell

Simulation and modeling have become practical diagnostic and predictive tools for large-scale industrial processes and equipment, allowing industrial scientists and engineers to perform "numerical experiments" during different stages of research, development and engineering design. For example, regeneration of catalysts in the Fluid Catalytic Cracking (FCC) process in different flow regimes can be simulated using a detailed CFD model combined with the simplified reaction kinetics. Results from such a model can help make informed decisions on the design and operating conditions for better process performance, improved energy efficiency and increased operating sustainability.

This presentation gives a brief overview of different approaches to gas-particle CFD as they apply to different FCC regenerator configurations. Flow patterns inside these vessels directly affect their performance and efficiency (of carbon burn and oxygen use) and serve as a qualitative guide to regenerator design improvement. "Numerical tracer" studies can provide a quantitative demonstration of the differences in residence time distributions of both catalyst and gas phase in different regenerator types. Moments of residence time distributions allow one to build ideal reactor networks to simulate the regenerator in conjunction with coke burn kinetics. Such a model shows the differences in coke combustion patterns inside the regenerator, demonstrates the effect of hardware upgrades on the regenerator performance, and ultimately serves as a design tool to assist process and equipment improvement efforts at Honeywell UOP.

Application of Numerical Modeling on Assessing Product Value Proposition and Enhancing Fundamental Understanding in Corning Incorporated

Authors:

Li, Y., Corning Inc.

Numerical modeling is now playing an important role in Corning. It has been used extensively to design new product, develop new process platforms, characterize the value propositions of new product, and enhance the fundamental understanding of the material and process. In this talk, we will give two examples about how we applied numerical modeling for Corning businesses.

In the first example, numerical modeling is used to characterize the value proposition of Gorilla hybrid laminate on improving the defrosting efficiency of vehicle windshield. The modeling results indicate that the Gorilla hybrid laminate shows 15% improvement over conventional counterparts in the standard test using a 0.5 mm frost layer. And the improvement is more pronounced in the tests with thin frost layers.

In the second example, numerical modeling is applied to enhance the fundamental understanding of the KGF transport inside cell clusters. The modeling results support the hypothesis that the KGF transport process is governed by the passive diffusion due to concentration gradient, and there is no noticeable active transport by the cells.

Membrane Organic Separations: An Overview

Authors:

Lively, R., Georgia Institute of Technology

The rapid increase in global industrialization necessitates technology shifts in energy production, manufacturing, and carbon management techniques. Large energy costs in refineries, power plants, and manufacturing facilities using traditional separation techniques are currently a major opportunity for innovation. Approximately 10% of global energy use can be attributed to separation processes, with the vast majority of separations being "thermal" in nature (e.g., distillation). Significant energy and cost savings can be realized using advanced separation techniques such as membranes and sorbents. One of the major barriers to acceptance of these techniques remains linking engineering materials to actual processes that are effective in the presence of aggressive industrial feeds (Sholl and Lively, 2016).

We created free-standing carbon molecular sieve membranes that translate the advantages of reverse osmosis for aqueous separations to the separation of organic liquids (Lively and Sholl, 2017). High-performance membranes derived from carbon materials have shown excellent chemical resistance, high molecular selectivity and fast mass transport across the membrane. Carbon molecular sieve (CMS) membranes with tailored ultramicropore and micropore dimensions show both high processability of polymeric membrane and the high selectivity of inorganic membranes. CMS membranes have been proven to be effective in a variety of gas separation processes such as olefin/paraffin separation, natural gas separation and air separations. However, the low fluxes observed in CMS hollow fibers (due to porous substructure collapse during pyrolysis) hinder scale-up of CMS membranes for industrial separation applications. The performance of carbon molecular sieve separation membranes, which exploit the effect of mass transport across a selective diffusion barrier to separate molecules can be improved by reducing the thickness of the membrane. We discuss methods to create hollow fiber membranes with thin CMS skin layers. We show that these membranes operate in "organic solvent reverse osmosis" (OSRO) separation modalities and purify p-xylene at room temperature without requiring any phase change (Koh et al., 2016).

References

"Seven chemical separations to change the world", DS Sholl, RP Lively, Nature 2016, 532, 435-437.

"From water to organics: revolutionizing membrane separations", RP Lively, DS Sholl, Nature Mater. 2017, 16, 276-279.

"Reverse osmosis molecular differentiation of organic liquids using carbon molecular sieve membranes," DY Koh, BA McCool, HW Deckman, RP Lively, Science 2016, 353(6301), 804-807.

Major Applications of OSN Membrane Technology in Pharmaceuticals and Fine Chemicals

2018 AIChE Process Development Symposium

Oak Brook, IL 5-7 June 2018

Kumar Abhinava, PhD
06-June-2018

EVONIK
POWER TO CREATE

Kumar Abhinava – Technical Director, OSN Applications

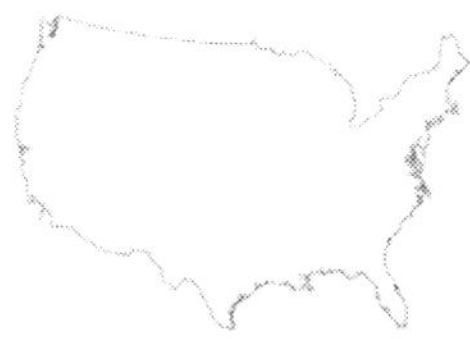

2011 -

EVONIK INDUSTRIES

2003 - 2011

Lilly

1989 - 2003

ICI ZENECA Avecia

1997 -	1997	1982 - 1989
IChemE	The University of Warwick	UMIST
Fellow	MBA	BSc; MSc; PhD Chemical Engineering

- **Over 28 yrs experience of Pharmaceuticals and Fine Chemicals Industries:-**
  - **Developing and commercializing NF and UF membrane processes**
  - **Process development, scale-up, optimization, troubleshooting,**
  - **Introduction of new process technologies**
  - **Expertise in several unit operations**

2

EVONIK
POWER TO CREATE

Outline of the Presentation

3

EVONIK
POWER TO CREATE

Who we are
Evonik at a glance

~13
Billion Euro sales in 2016

80%
Of turnover gained from leading market positions

172
Sites

>36,000
Employees in over 100 countries

2.165 billion Euro
EBITDA 2016

~500
R&D projects

4

EVONIK
POWER TO CREATE

Where we can be found
Evonik operates worldwide

Sales in 2016

5

EVONIK
POWER TO CREATE

Where we can be found
Our profile in the region North America

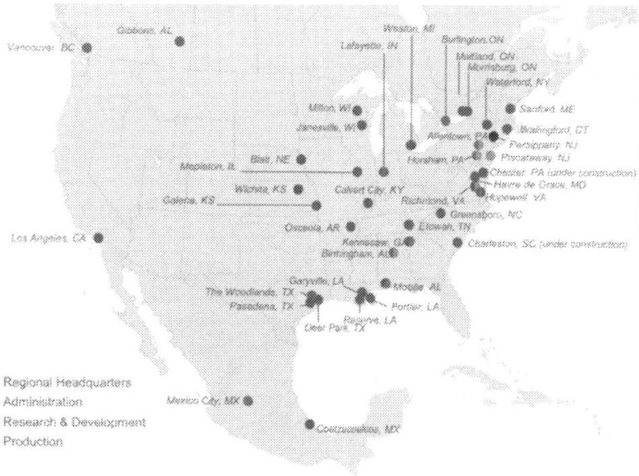

4,660 Employees	€ 2,669 M Sales
39 Sites	Home of the Project House Medical Devices Special Fact

6

EVONIK
POWER TO CREATE

Resource Efficiency
Strategic concentration and operative independence

Resource Efficiency

Environmentally friendly and energy-efficient systems as solutions for several industries

Business Lines

Coating Additives
Additives and special resins for coatings, colorants, and printing inks

Catalysts
Catalysts for synthesizing pharmaceuticals, fine and industrial chemicals

Oil Additives
Methacrylate-based polymers for the lubrication and fuel industries

Silanes
Chlorosilanes, functional silanes, and rubber silanes as additives or raw materials

Active Oxygens
Hydrogen peroxide and peracetic acid for many industrial applications

Coating & Adhesive Resins
Specialty polymers for coatings, adhesives and sealants

Silica
Silicas, specialty oxides and dispersions, as well as matting agents

Crosslinkers
Isophorone and isophorone derivatives for efficient coatings and resins

High Performance Polymers
The expert for tailored solutions for high-performance polymers

Product Lines

Granules & Compounds
- Automotive
- Oil & Gas
- Consumer Goods
- Industry

Fibers, Membranes & Specialties
Fibers
Membranes: Gas, OSN
Medical: Devices, Implant
Specialties

Engineering Products
Powders, Foams
Composite

7

EVONIK
POWER TO CREATE

Membrane products from Fibres, Membranes and Specialties

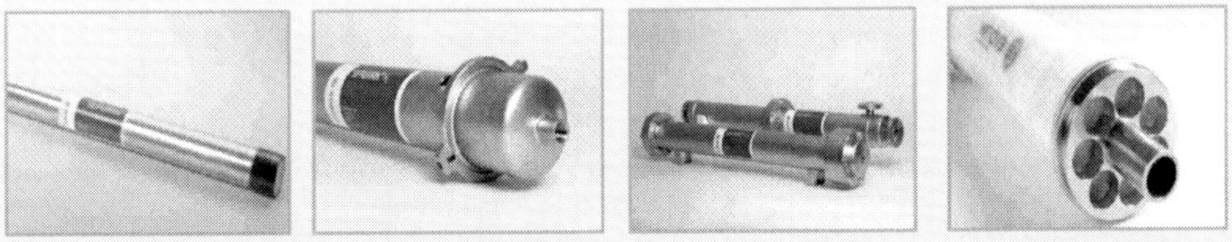

Sepuran ® Green modules for Biogas Separation	Sepuran ® N_2 modules for nitrogen separation	Sepuran ® Noble modules for helium separation	Puramem ® and Duramem ® modules for OSN [1]
Market leading biogas upgrading technology	Lowest total cost of ownership for nitrogen generation	Highly efficient helium recovery	High separation performance in polar and apolar organic solvents

8

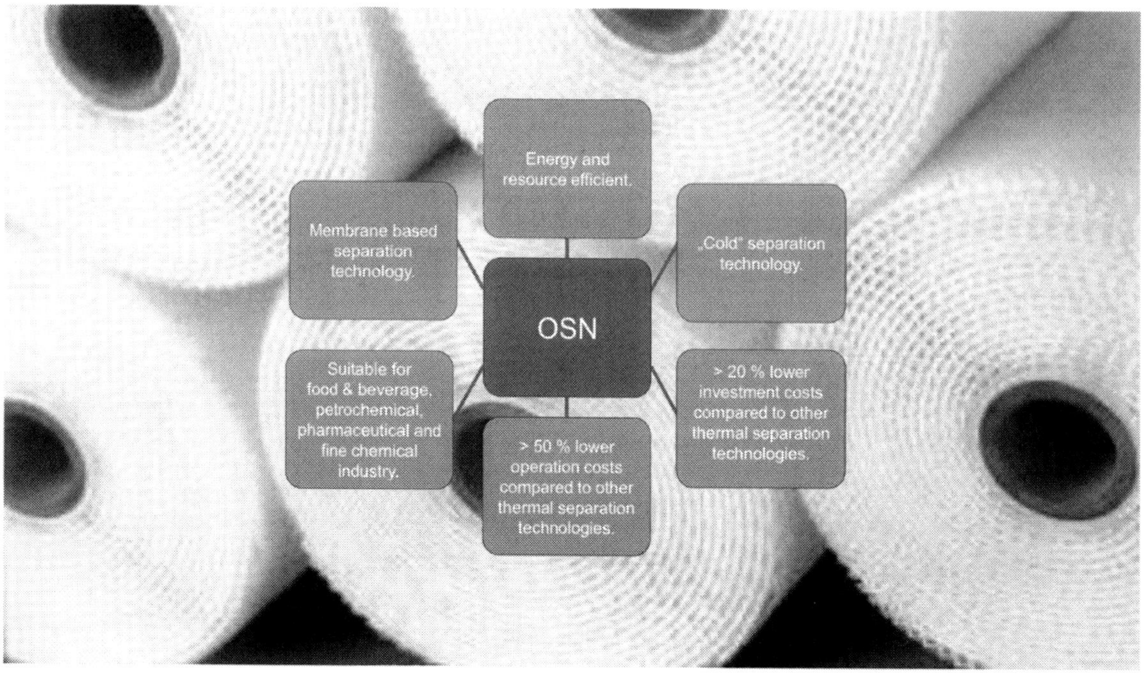

OSN is a disruptive molecular separation technology

What is Organic Solvent Nanofiltration (OSN)?

- It is a scalable molecular separation technology based on membranes
- It separates molecules dissolved in an organic solvent (mixture)
- It is an emerging technology for energy & material efficient separations
- ***Evonik's OSN Membrane Technology USP: Stability in a wide range of organic solvents***

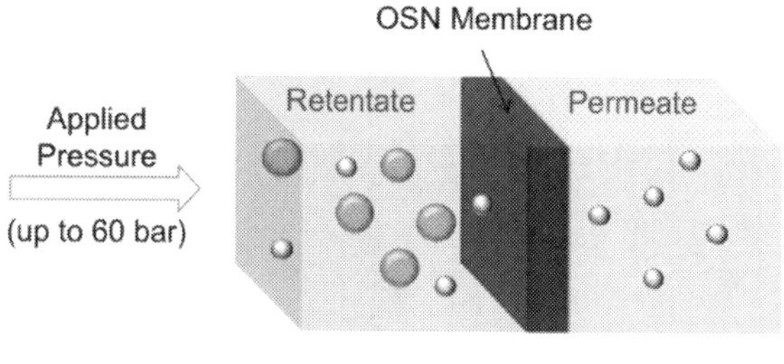

10

Spiral Wound Module Flow Schematic

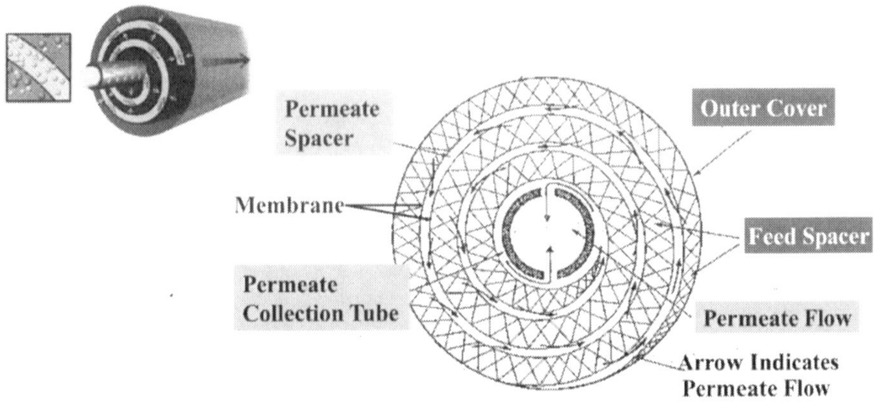

OSN Commercialized Membranes

	DuraMem (for polar solvent systems)	PuraMem (for non-polar solvent systems)	Technology
Generation 1	DM 150 (T1) DM 200 (T1) DM 200 (T2) DM 300 (T1) DM 500 (T1) DM 900 (T1)	PM 280	Phase-Inversion Asymmetric
Generation 2		PM S600	Thin Film Composite Dip Coated (no solvent)
Generation 3		PM Selective (350) PM Performance (400) PM Flux (450)	Thin Film Composite Dip Coated (with Solvent) (thinner than above)

PuraMem and DuraMem: General Observations

PuraMem Gen 3

- Industries: Bulk Chemicals; Green Chemistry; Biodiesel; Oil & Gas; Petrochemicals
- Membranes: PuraMem Flux, PuraMem Performance and PuraMem Selective. For one application, permeate flux rates were similar at high pressures with best rejections for PuraMem Selective
- Commercialization: Time to commercialization is long due to novel processes (e.g. green chemistry) and/or novel applications (e.g. Biodiesel)
- Volumes: Module counts required are in hundreds
- Pricing: Projects sensitive to pricing
- Custom Membrane Development: Request from several large companies for membranes that have a 'tighter' cut-off than PuraMem Selective membrane

DuraMem

- Industries: Pharmaceuticals ("Small Molecules" synthesis); Fine Chemicals (Natural Oils, Essential Oils, Alternative Biocrops etc., Electronics). Low Volume; High Value
- Membranes: DuraMem 500 and DuraMem 200 are particularly popular. Also, DuraMem is considered a consumable with service life expectation between 6-12 months.
- Commercialization: Time to commercialization is relatively short
- Volumes: Module counts required are low (Equivalent of 2-4 x 8" modules is typical)
- Pricing: Not particularly sensitive to pricing. Performance is key

13

EVONIK
POWER TO CREATE

Application Development

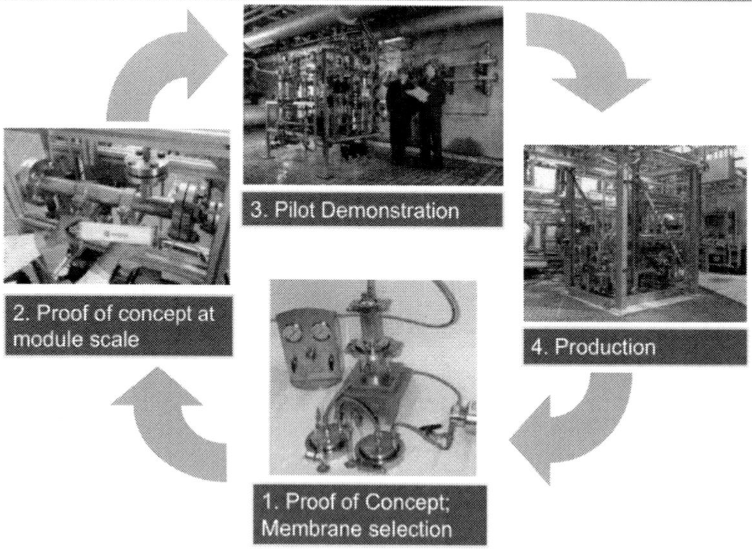

14

EVONIK
POWER TO CREATE

Fine Chemicals OSN Application Areas

Generalized Current Process Scheme for Natural Oils, Essential Oils, Alternative Biotech Crop Extraction

Process Scheme with OSN

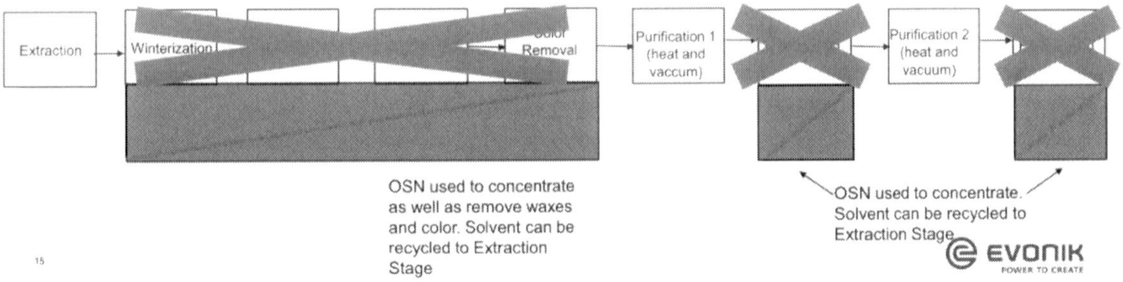

Fine Chemicals OSN Applications Areas

- 1. Simplification of Process e.g. remove winterization, filtration, de-colorization unit operations
  - CAPEX Savings
  - OPEX Savings
  - Yield and Quality Benefits
- 2. Non-thermal Separations
  - Yield and Quality Benefits
- 3. Non-Thermal Concentrations
  - Yield and Quality Benefits
  - Reduction in OPEX (reduced energy costs)
- 4. Unique Separation Characteristics
  - New Products (e.g. Fragrance / Flavor profile)

Pharmaceutical Manufacturing – OSN Applicable Steps / Unit Operations

Each box in the OSN applicable region consists of following unit operations. (Milling is only done for API)

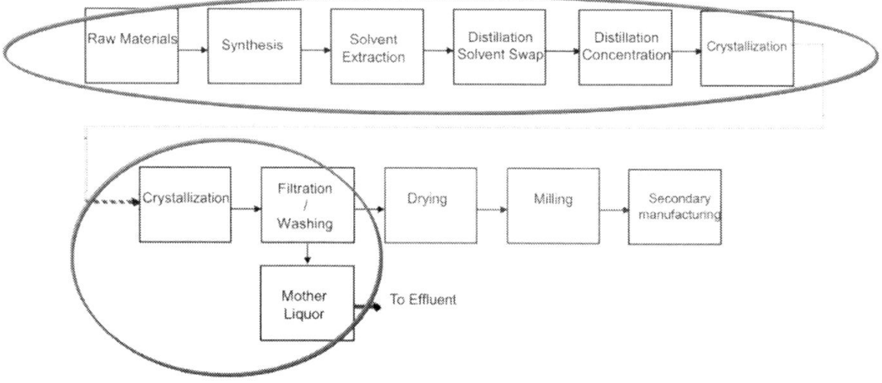

17

EVONIK POWER TO CREATE

OSN Applications Areas

OSN APPLICATIONS

- Reactions
- Extractions – API in Waste Stream
- Distillation – Solvent Swap
- Distillation – Concentration
- Crystallization - API in M-L

- 1. Removal of impurities (smaller MW than product)
- 2. Removal of impurities (larger MW than product)
- 3. Drive equilibrium reactions forward by removal of unwanted by-product
- 4. Drive towards use of stoichoimetric amounts of starting materials
- 5. Increase concentration (if it helps in achieving more complete reaction)
- 6. Solvent swap
- 7. Catalyst recovery and recycle
- *Above uses of OSN membranes in 'Reactions' can allow:-*
- 1. Continuous processes to be developed more readily
- 2. Telescoping of processes (i.e. avoid isolations and drying) – thus increasing overall yields
- 3. Achieve cleaner and greener chemistry

18

EVONIK POWER TO CREATE

Johnson & Johnson: API Purification & Solvent Recovery Application

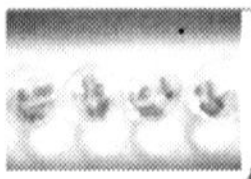

➤ OBJECTIVE: To Purify C by separating from D

$$A \xrightarrow{B} C + D$$

Material	Function	MW (g.mol^{-1})
A	Reactant	700
B	Catalyst	600
C	API Intermediate	680
D	Oligomers of C (IMPURITY)	≥ 1400
THF	Solvent	72

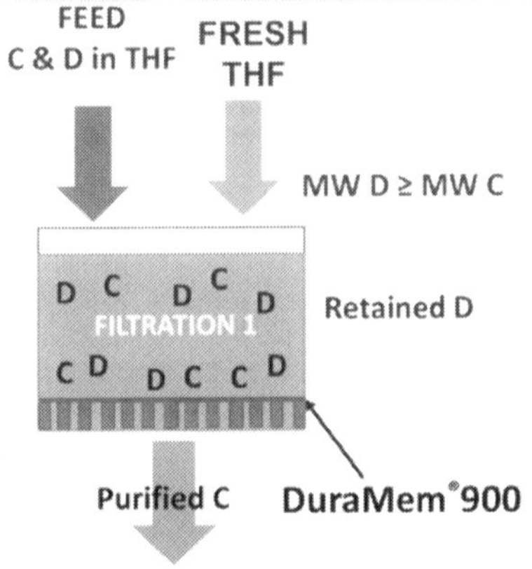

OSN process successfully:
- Recover ≥ ***99%*** of Product C
- Generates product containing less than 2.5% of Impurity D (≤ 3.0wt% ***threshold***)

19

EVONIK CONFIDENTIAL

BUT, **requires** 380 L **of solvent to purify 1 kg of Product C!!**

EVONIK POWER TO CREATE

Seite | 19

Johnson & Johnson: API Purification & Solvent Recovery Application

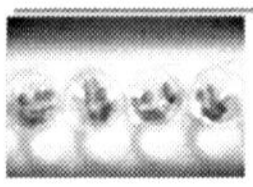

Process	Function
Filtration 1	Purification
Filtration 2	Solvent Recovery

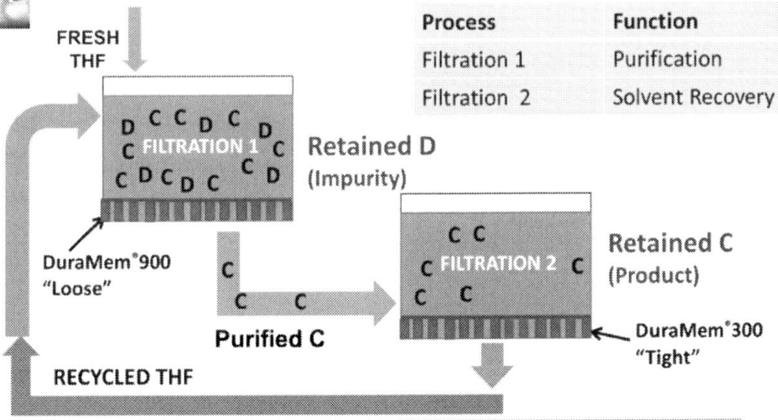

With recycle, **≤ 38 L of solvents required to purify 1 kg of Material C, which is** 10 X less **solvent!!**

20

EVONIK POWER TO CREATE

Homogeneous Catalyst Recovery (HCR)

Feed: Reaction mass following synthesis

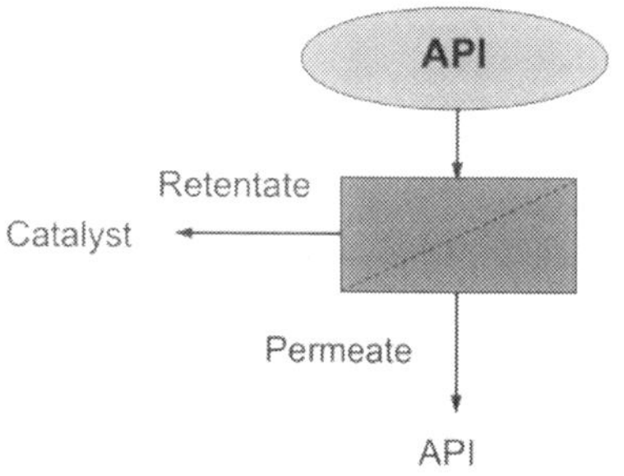

Results:-

- DM500 gave god rejection and flux

Application:-

Homogeneous Catalyst Recovery (HCR)

MW: API ~ 300; Catalyst ~ 800

Solvent: THF

Why choose OSN:-

1. HC is expensive
2. Current technology not cost effective for HC recovery
3. Current technology does not guarantee 100% removal

Trials carried out:-

MCXF: DM500

21

EVONIK POWER TO CREATE

Homogeneous Catalyst Recovery: Results

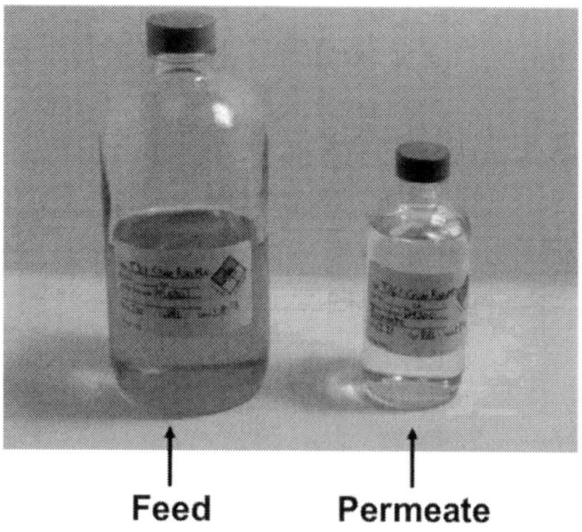

- **Visually, separation appears to be successful.**
- **Analysis by customer shows excellent results**

22

EVONIK POWER TO CREATE

GSK Production Plant: Improved Process Yield

PROBLEM

- Acetone waste solution containing ~1 wt% of valuable
- final API (~400 $g.mol^{-1}$) generated in final step of API production
- Disposal of this stream constitutes significant yield loss
- API highly temperature sensitive – rapidly decomposes above ambient temperature
- Evaporation/distillation not possible on this stream
- API must be concentrated to >10wt% for recycle

23

GSK Example: Yield Improvement. API Recovery

API Production Process

- Batch-production process
- 20 m3/d waste stream containing API
- Value of API > 500k €

Thermal separation

Energy costs: 30k €/a

Investment costs: 300k €

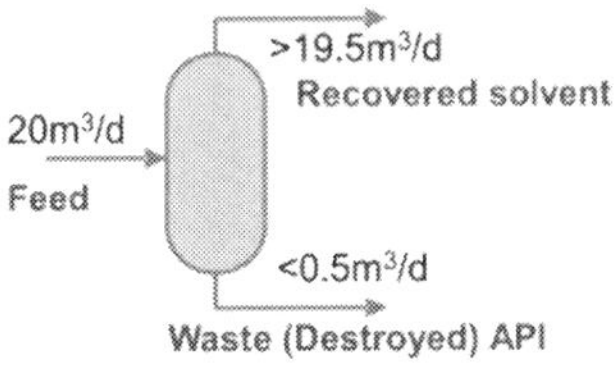

Membrane process:

Membrane costs: 25k €/a

Investment costs: 300k Mio. €

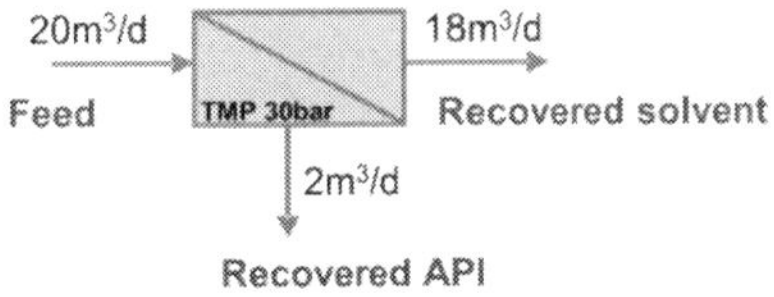

24

OSN → **Comparable CapEx and OpEx** } Revenue *Increase* **> 500k €/a**

API Recovery from M/L Application: Process Schemes

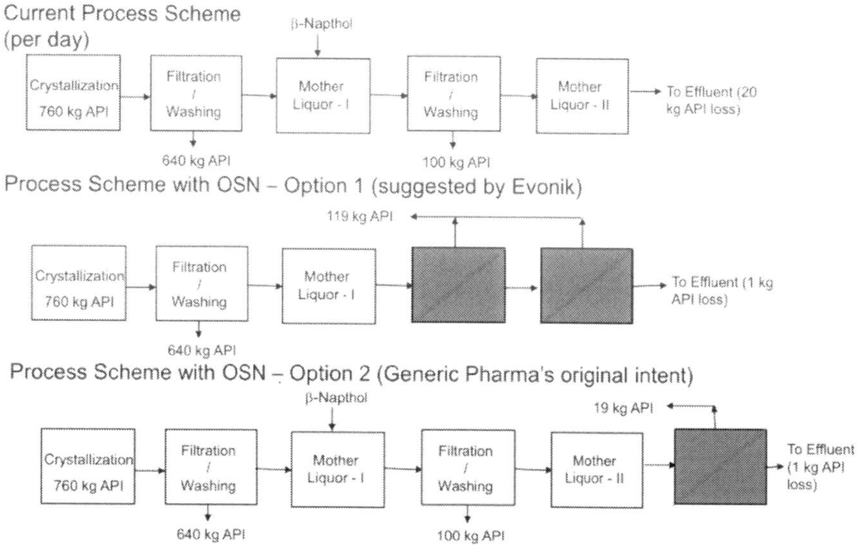

25

EVONIK
POWER TO CREATE

AstraZeneca Example: Improvement in Product Quality (Color Removal)

- Separation of coloured impurity from API at Astra Zeneca by OSN *

(3a) Starting material containing high MW coloured compound

(3b) Coloured impurity difficult to remove via extraction

(3c) Product after OSN purification. Nice white powder!

august 2006 tce 31

Nanofilter solution so that solvent + API passes through membrane and impurity is retained

26

See The Chemical Engineer, August 2006

EVONIK
POWER TO CREATE

DMSO Removal (Solvent Swap with Acetonitrile) to Improve Crystallization

Feed: Intermediate in DMSO / Acetonitrile

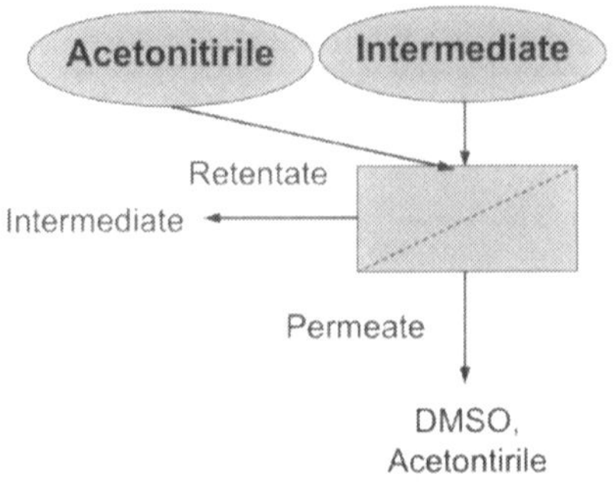

<u>Application:-</u>

DMSO removal by solvent swap (diafiltration)

MW: Intermediate > 200

Solvent: DMSO / Acetonitrile

<u>Why choose OSN:-</u>

1. DMSO > 0.5% w/w may cause issues in unit operations such as crystallization, purification and drying
2. Alternative DMSO removal processese such as chromatography could be expensive and difficult to operate and scale-up

<u>Membrane Chosen:-</u>

DuraMem 200

27

EVONIK POWER TO CREATE

OSN Purification Application: Generic Pharma

Feed: Fermentation Broth (Toluene)

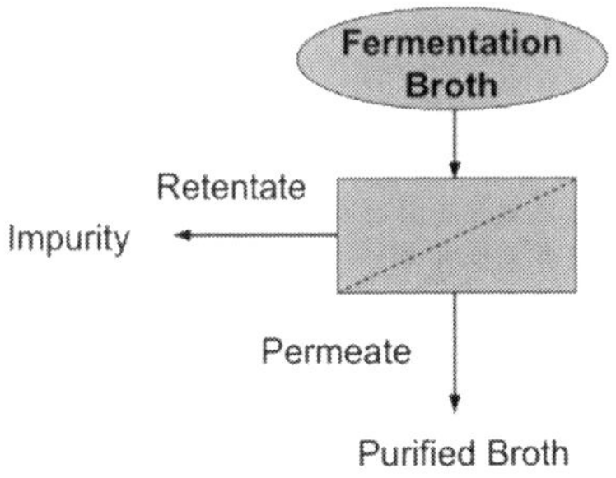

<u>Application:-</u>

Removal of high MW impurity

MW: Impurity: Large; Solvent: Toluene

<u>Objectives:-</u>

1. Increase product yield by removal of high MW impurity that impacts the yield during the down-stream chemistry

<u>Trials carried out:-</u>

1. MCXF with PMS600
2. BTU Module with PMS600

Benefits of OSN Membrane Technology :-

- Increased yield
- Longer life of hydrogenation catalyst (cost savings)
- Recovery of heterogeneous catalyst possible (cost savings)

28

EVONIK POWER TO CREATE

OSN Purification Application: Generic Pharma

Current Process Scheme

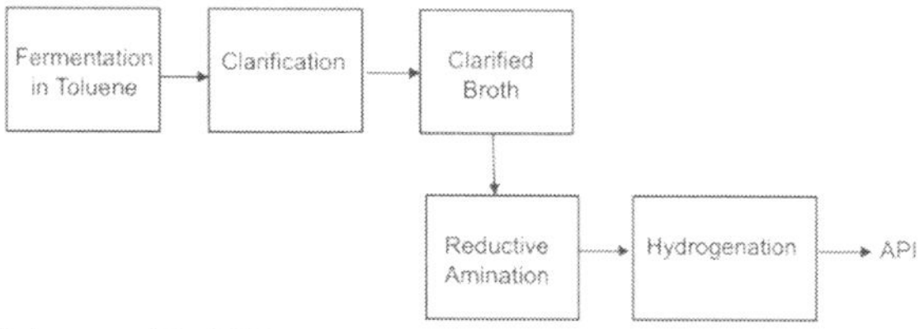

Process Scheme with OSN

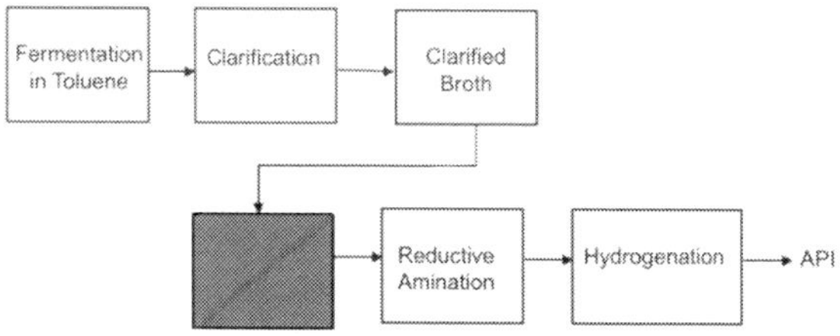

29

EVONIK POWER TO CREATE

Other Pharma Applications

- API and solvent recovery from Chromatography Steps (Chromatography is usually used instead of or downstream of Crystallization to achieve extremely high level of purity)

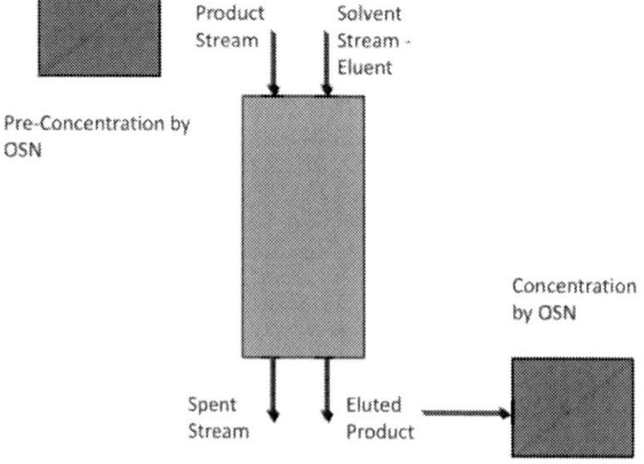

30

EVONIK POWER TO CREATE

Purification and Concentration: Process Scheme

Current Process Scheme

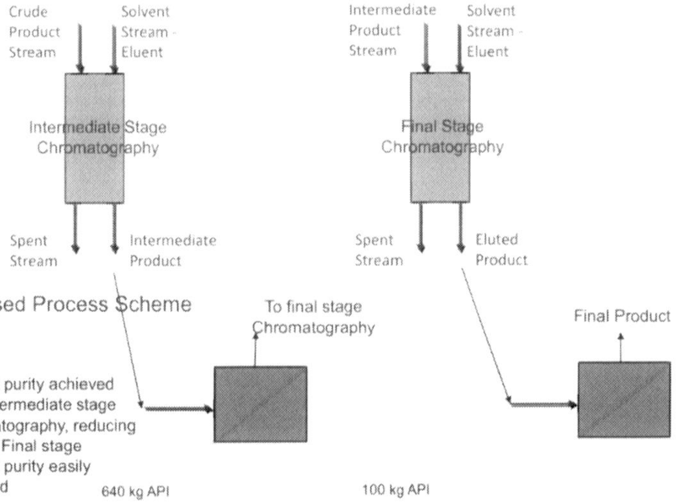

Proposed Process Scheme

Benefits:-

1. Product purity achieved from Intermediate stage Chromatography, reducing load on Final stage
2. Product purity easily achieved

EVONIK POWER TO CREATE

Seite | 31

31

Summary

Evonik have strong know-how in the field of manufacturing of and processing with OSN membranes

Evonik recently launched a new generation of TFC (Thin Film Composite) membranes with superior performance in June 2016

Evonik is long standing user and supplier of OSN technologies and a strong partner for many companies in multiple industry segments, espeically pharmaceuticals

32

EVONIK POWER TO CREATE

EVONIK
POWER TO CREATE

Exploring Membrane Separations in Hydrocarbon Processing

Authors:

McCool, B., ExxonMobil

ExxonMobil has been interested in the use membranes for both gas and liquid separations for at least the last 25 years. In that time we have investigated numerous applications with and worked with inorganic membranes of all types. That work resulted in over 100 granted US patents and over 30 publications in the peer reviewed literature. This presentation will provide an overview of the work from the past 25 years – including key findings in our continued progression toward understanding molecular transport and materials engineering aspects of inorganic membrane separations. Specific examples will be presented on zeolite membranes for the separations of CO2/CH4 and xylenes as well as hybrid systems for aromatics/non-aromatics pervaporation. The second half of the presentation will focus on our on-going work in the separation of complex hydrocarbon mixtures using polymeric and inorganic membranes. We will cover our efforts in novel materials development, compositional modeling for separation products in complex mixtures, and process concept development for nano-filtration and hydrocarbon reverse osmosis applications.

1

Process development for production of Tier 3 ultra-low sulfur gasoline.

Bringing an old process into the 21st Century

Hoekstra Trading LLC

2

Gasoline desulfurization

- **The business case**
- Studying the detailed chemistry of the process
- Commercial field tests

Hoekstra Trading LLC

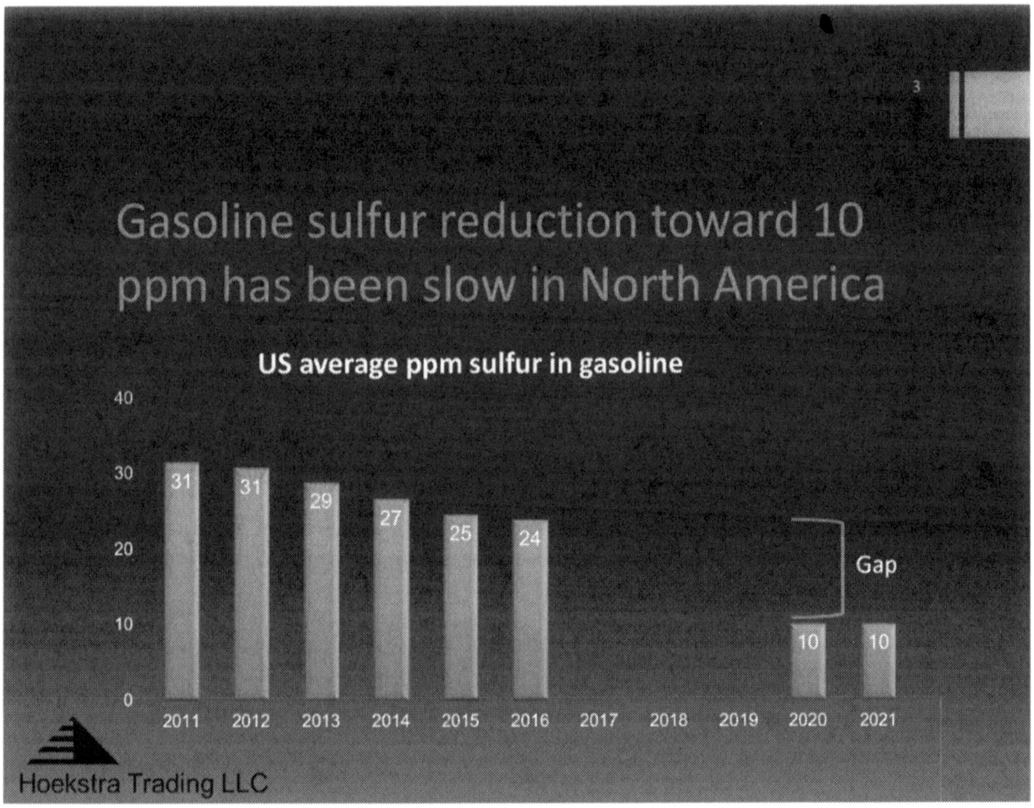
3
Gasoline sulfur reduction toward 10 ppm has been slow in North America
US average ppm sulfur in gasoline
40
30
20
10
0
31
31
29
27
25
24
Gap
10
10
2011
2012
2013
2014
2015
2016
2017
2018
2019
2020
2021
Hoekstra Trading LLC

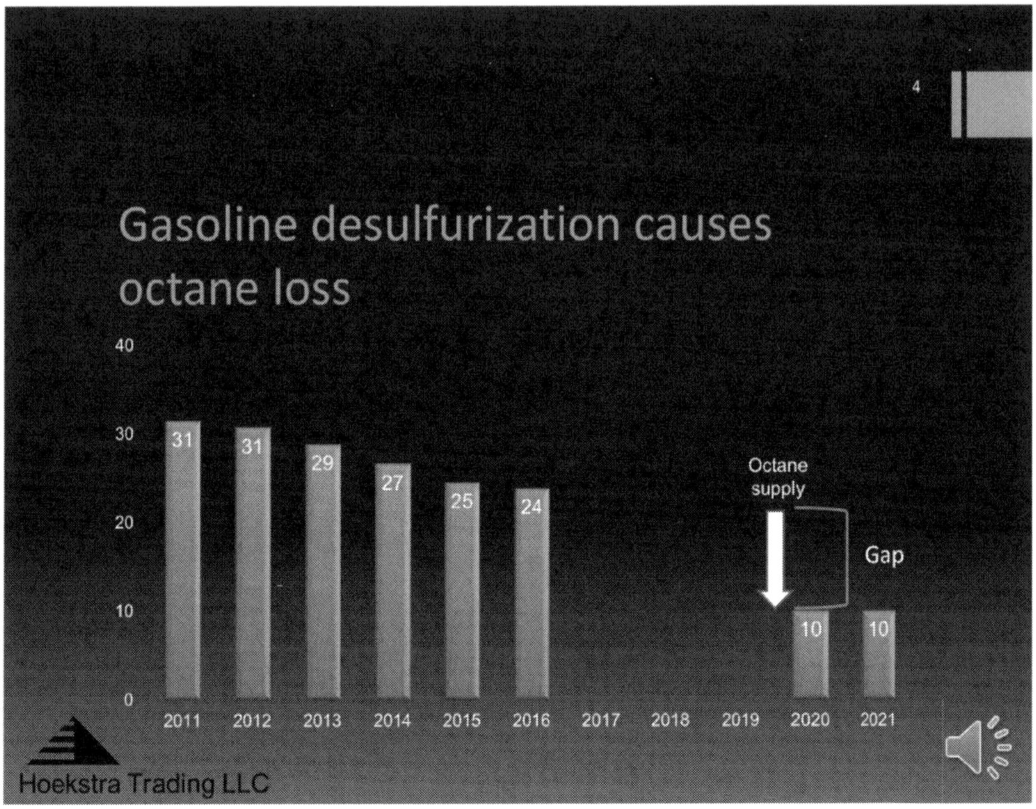
4
Gasoline desulfurization causes octane loss
40
30
20
10
0
31
31
29
27
25
24
Octane supply
Gap
10
10
2011
2012
2013
2014
2015
2016
2017
2018
2019
2020
2021
Hoekstra Trading LLC

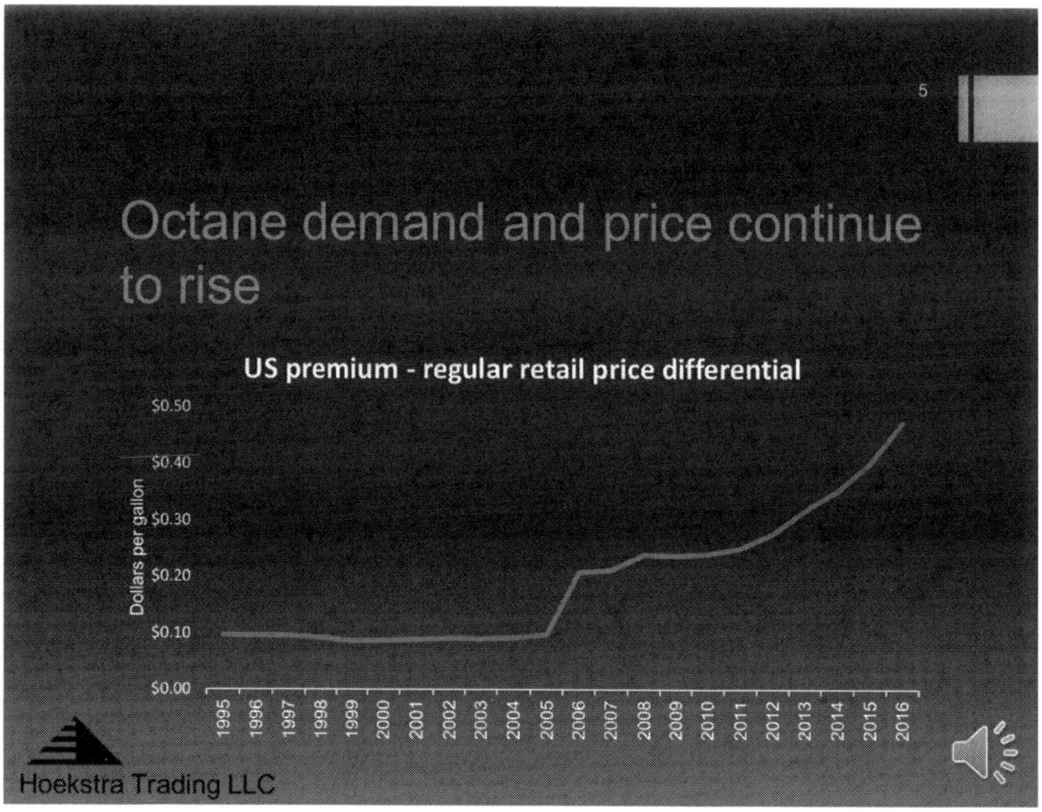
5
Octane demand and price continue to rise
US premium - regular retail price differential
$0.50
$0.40
$0.30
$0.20
$0.10
$0.00
Dollars per gallon
1995
1996
1997
1998
1999
2000
2001
2002
2003
2004
2005
2006
2007
2008
2009
2010
2011
2012
2013
2014
2015
2016
Hoekstra Trading LLC

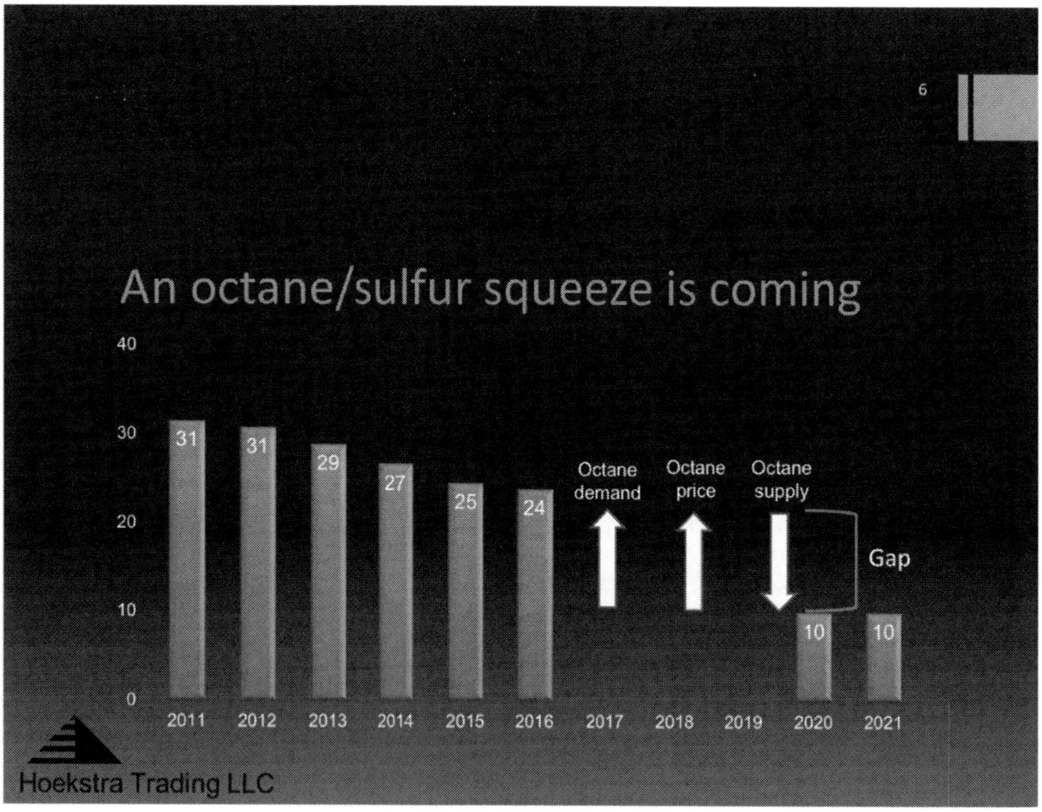
6
An octane/sulfur squeeze is coming
40
30
20
10
0
31
31
29
27
25
24
Octane demand
Octane price
Octane supply
Gap
10
10
2011
2012
2013
2014
2015
2016
2017
2018
2019
2020
2021
Hoekstra Trading LLC

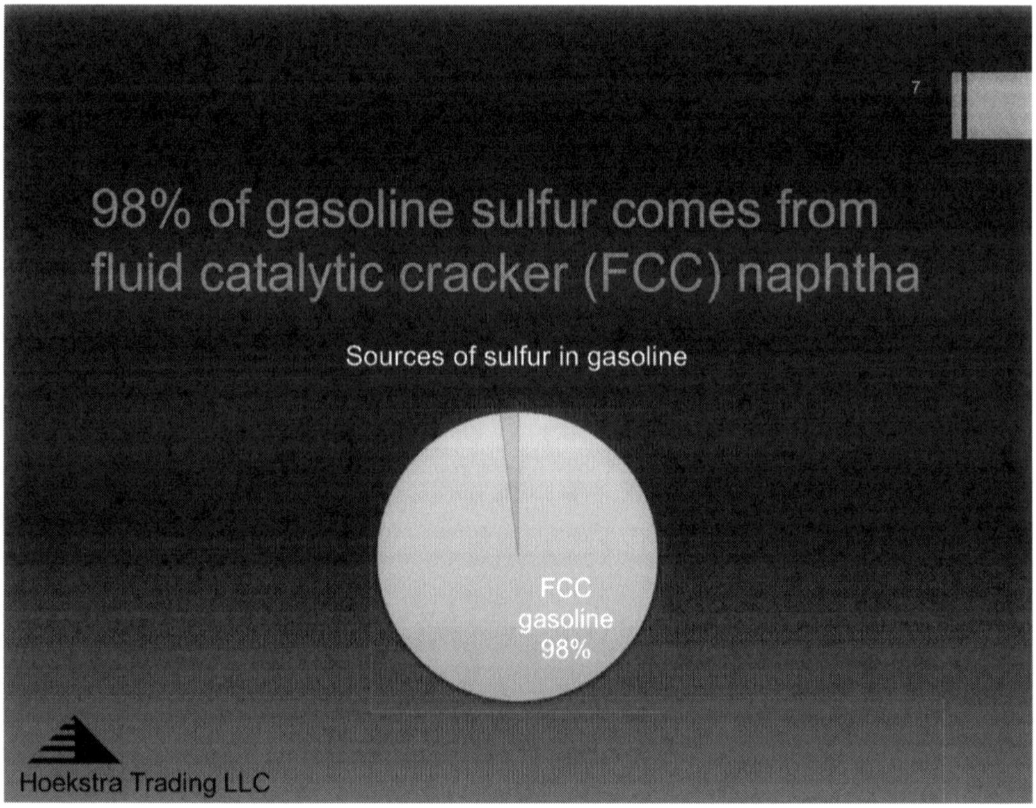
7
98% of gasoline sulfur comes from fluid catalytic cracker (FCC) naphtha
Sources of sulfur in gasoline
FCC gasoline 98%
Hoekstra Trading LLC

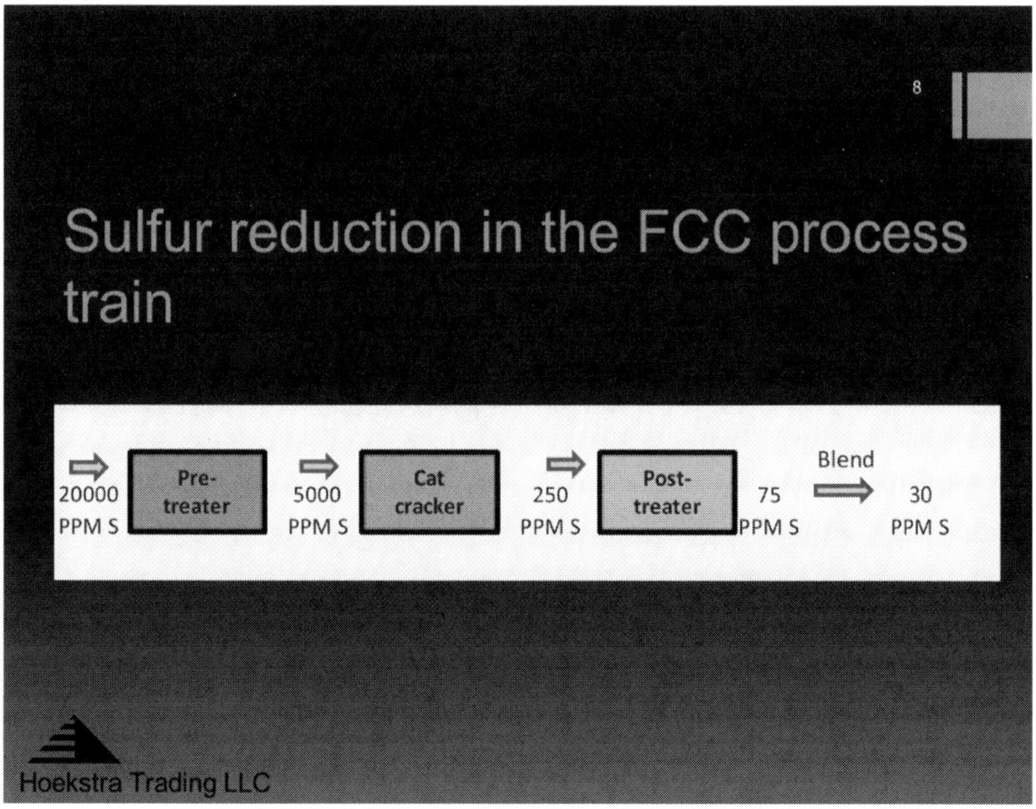
8
Sulfur reduction in the FCC process train
20000 PPM S
Pre-treater
5000 PPM S
Cat cracker
250 PPM S
Post-treater
75 PPM S
Blend
30 PPM S
Hoekstra Trading LLC

9

FCC gasoline post-treating causes octane loss

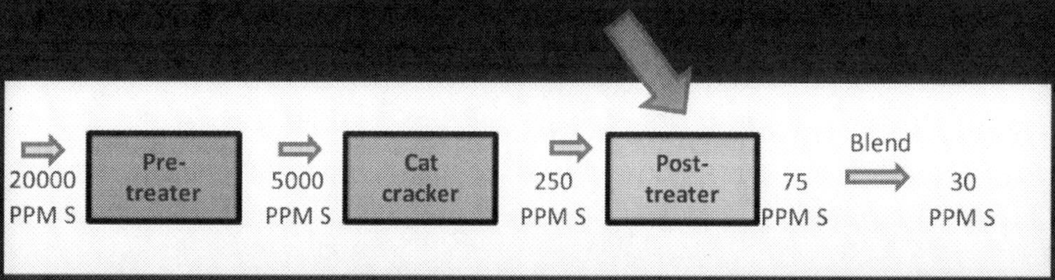

Hoekstra Trading LLC

10

Gasoline desulfurization

- The business case
- **Studying the detailed chemistry of post-treating**
- Commercial field tests

Hoekstra Trading LLC

FCC gasoline post-treating

- Goal: desulfurize FCC gasoline with minimum octane loss
- Uses conventional fixed bed naphtha hydrotreating
- Uses selective catalysts to minimize olefin saturation

Hoekstra Trading LLC

FCC gasoline post-treating

- Got off the ground in 1999
- There are now 300 units in the world
- Most are Scanfining (Exxon) or Prime G (Axens) units
- Revamps and new builds are ongoing

Hoekstra Trading LLC

Pilot plant testing

C Solutions LTD.
Thessaloniki, Greece

Hoekstra Trading LLC

The Performance curve

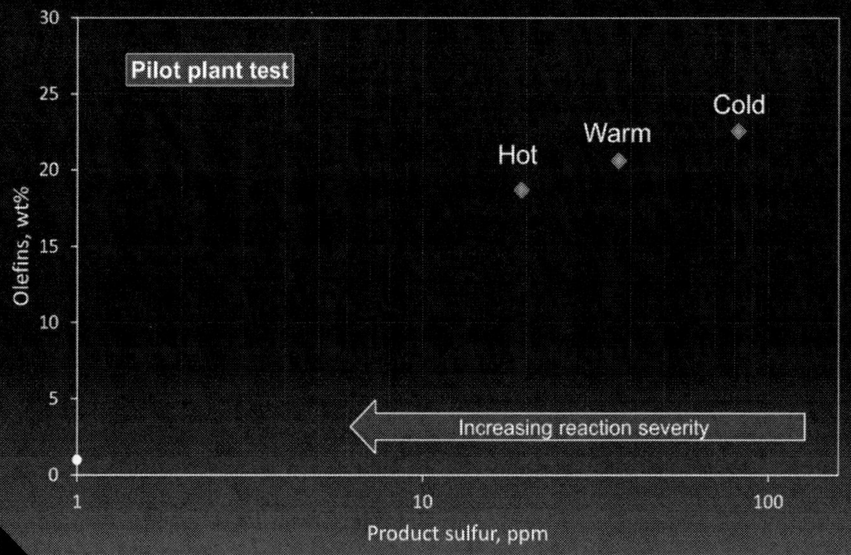

Hoekstra Trading LLC

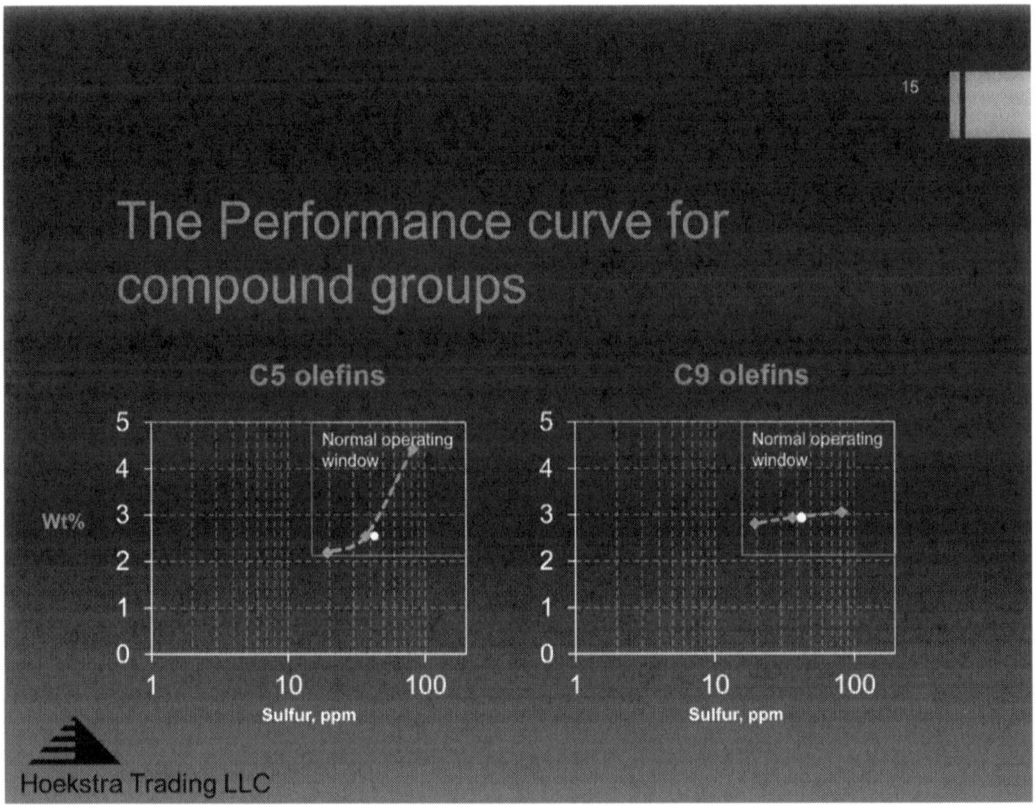
15
The Performance curve for compound groups
C5 olefins
C9 olefins
Normal operating window
Normal operating window
Wt%
5
4
3
2
1
0
1
10
100
Sulfur, ppm
Sulfur, ppm
Hoekstra Trading LLC

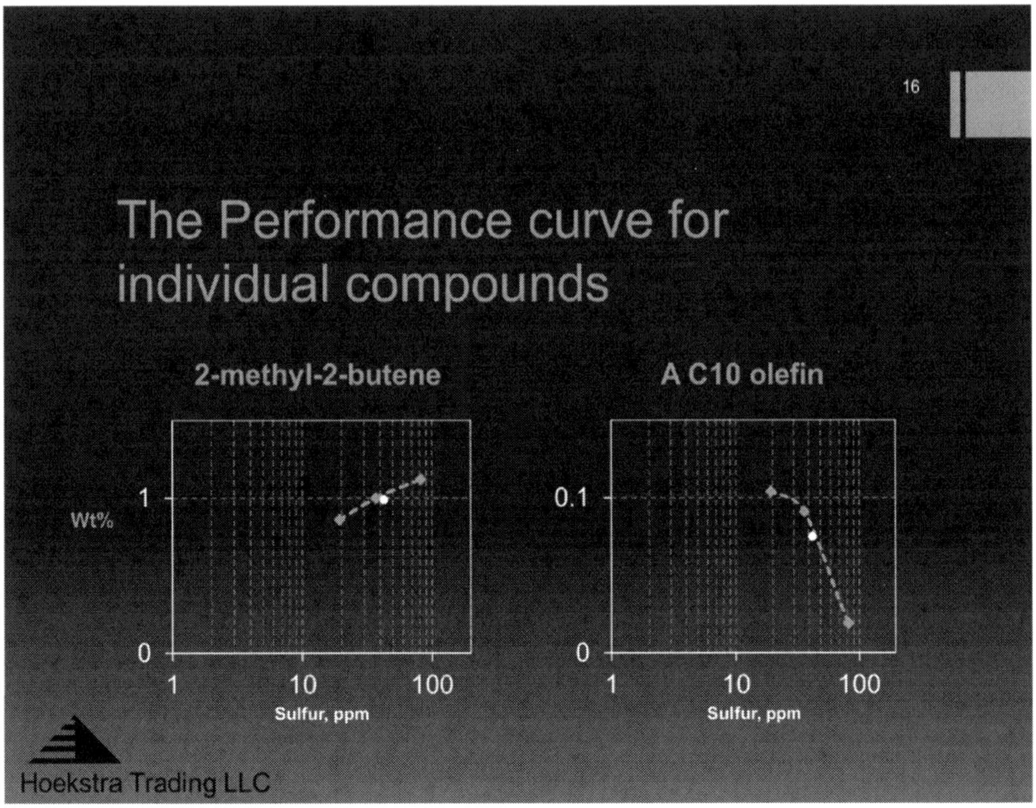
16
The Performance curve for individual compounds
2-methyl-2-butene
A C10 olefin
Wt%
1
0
0.1
0
1
10
100
Sulfur, ppm
Sulfur, ppm
Hoekstra Trading LLC

Pushing the process to make 10 ppm S

17

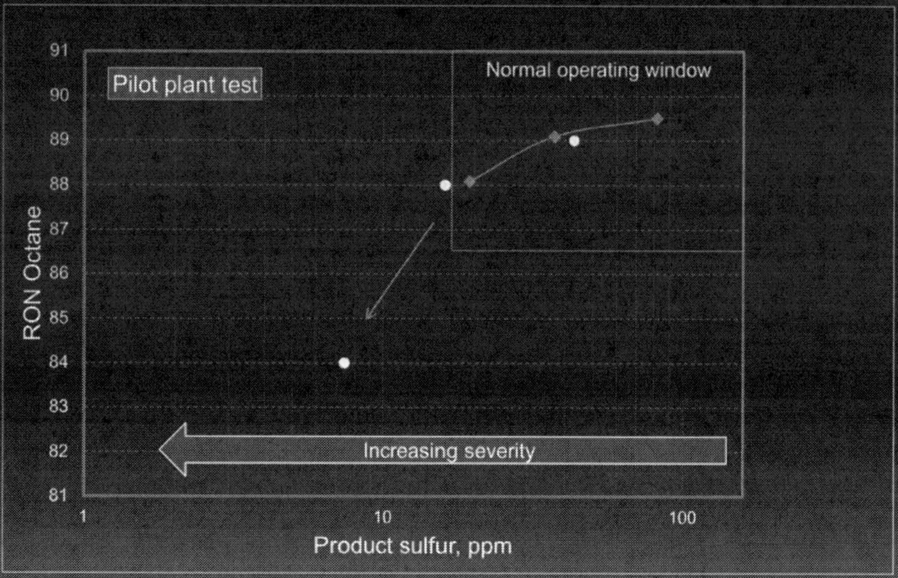

Hoekstra Trading LLC

18

Gasoline desulfurization

- The business case
- Studying the detailed chemistry of the process
- **Commercial field tests**

Hoekstra Trading LLC

Olefin conversion by carbon number

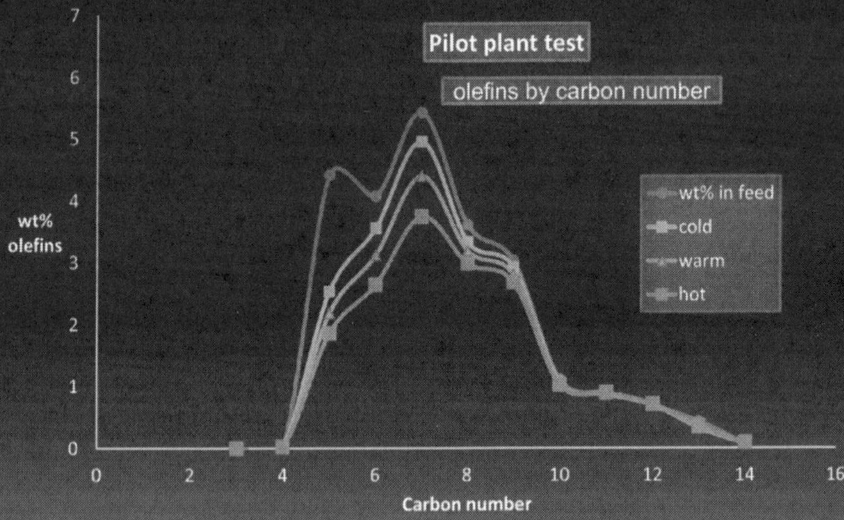

Hoekstra Trading LLC

Hoekstra "blood tests" on commercial units

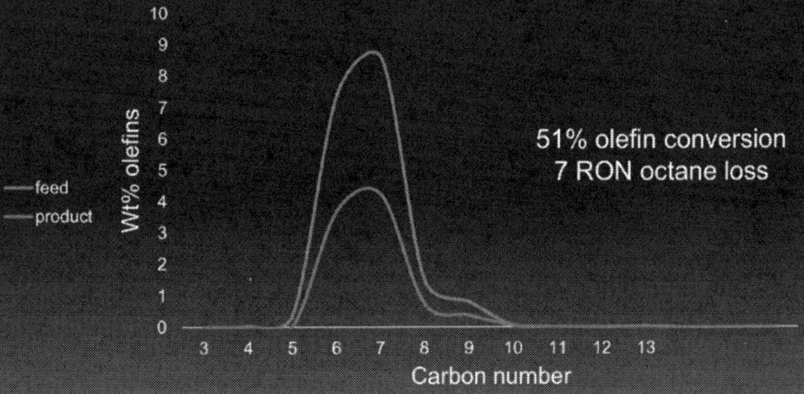

Hoekstra Trading LLC

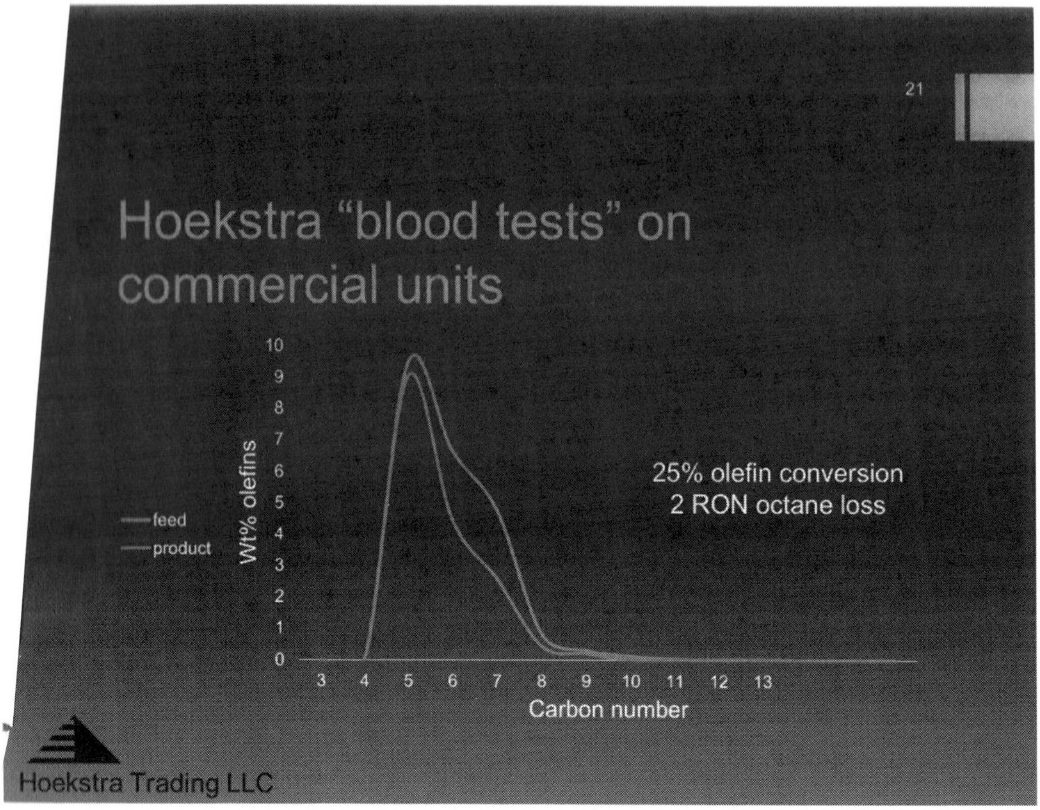
21
Hoekstra "blood tests" on commercial units
Wt% olefins
10
9
8
7
6
5
4
3
2
1
0
3 4 5 6 7 8 9 10 11 12 13
Carbon number
feed
product
25% olefin conversion
2 RON octane loss
Hoekstra Trading LLC

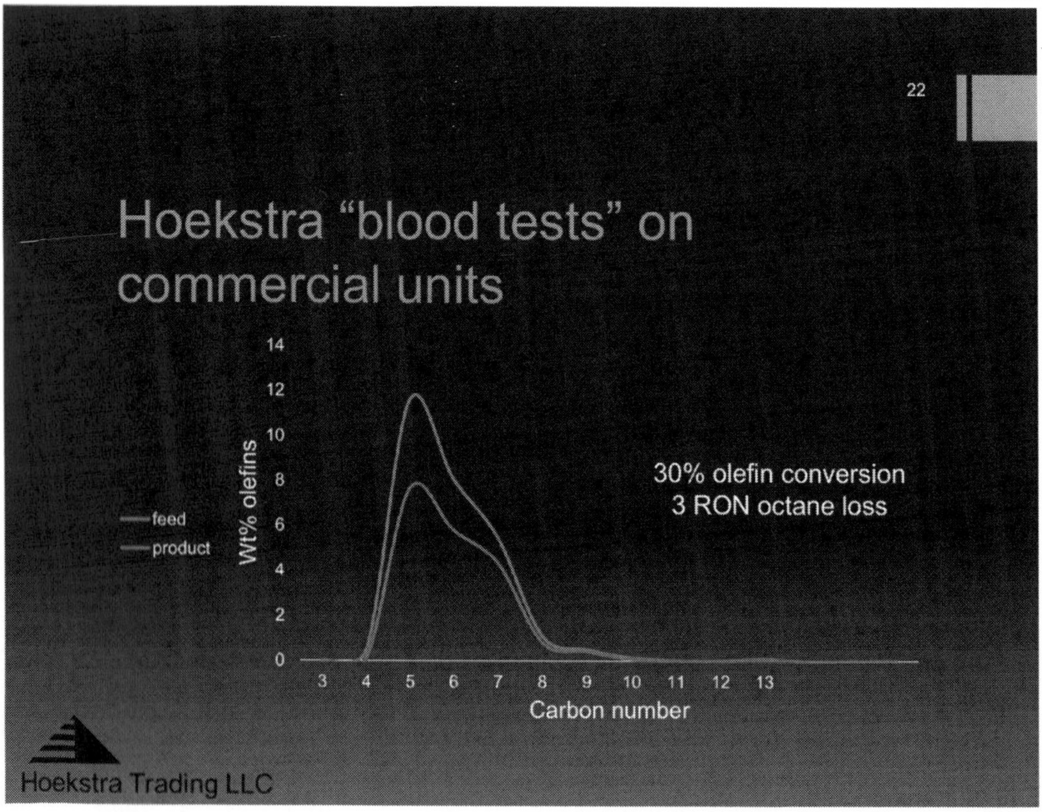
22
Hoekstra "blood tests" on commercial units
Wt% olefins
14
12
10
8
6
4
2
0
3 4 5 6 7 8 9 10 11 12 13
Carbon number
feed
product
30% olefin conversion
3 RON octane loss
Hoekstra Trading LLC

23

Conclusions

- An octane/sulfur squeeze is coming
- High severity post-treating accelerates octane loss
- Modern analytical methods point to the specific causes
- Capital investments will be needed to mitigate octane loss

Hoekstra Trading LLC

24

Thanks for your attention!

- George Hoekstra

- george.hoekstra@hoekstratrading.com
- http://hoekstratrading.com
- +1 630 330-8159

Hoekstra Trading LLC

Designing Extraction Columns for Fractional Liquid-Liquid Extraction

Authors:

Glatz, D., Koch Modular

Often times a simple liquid-liquid extraction process (single counter current solvent) will not work effectively for certain separations due to mutual extraction of a number of components into the solvent phase. Many of these separations are ideal candidates for an under utilized technique known as fractional extraction. Like simple extraction, the best method for designing an extraction column for a fractional extraction process is via pilot plant testing followed by scale up from the pilot data. This report demonstrates this method of design using two different separation applications. One application involves the separation of a mixed polymer feed. The second one is a heavy naphtha feed that is separated into polar aromatics and non-aromatic impurities. Pilot plant equipment set-ups are presented along with the test procedures used to optimize the performance for both applications. Scale up to the production columns are discussed.

Successful Scale-Up and Commercialization of a Breakthrough Solid Acid Alkylation Technology: The AlkyClean® Process

June 2018

MCDERMOTT TECHNOLOGY

LUMMUS TECHNOLOGY

FORWARD LOOKING STATEMENT

McDermott cautions that statements in this presentation which are forward-looking, and provide other than historical information, involve risks, contingencies and uncertainties that may impact actual results of operations. These forward-looking statements include, among other things, statements about anticipated cost and revenue synergies, accretion, best-in-class operations, opportunities to capture additional value from market trends, maintenance of a consistent customer approach to pricing, safety and transition issues, free cash flow, plans to de-lever and accretion. Although we believe that the expectations reflected in those forward-looking statements are reasonable, we can give no assurance that those expectations will prove to have been correct. Those statements are made by using various underlying assumptions and are subject to numerous risks, contingencies and uncertainties, including, among others: the possibility that the expected synergies from the recently completed combination will not be realized, or will not be realized within the expected time period; difficulties related to the integration of the two companies; disruption from the combination making it more difficult to maintain relationships with customers, employees, regulators or suppliers; the diversion of management time and attention related to integration matters; adverse changes in the markets in which the company operates; the inability to execute on contracts in backlog successfully; changes in project design or schedules; the availability of qualified personnel; changes in the terms; scope or timing of contracts; contract cancellations; change orders and other modifications and actions by customers and other business counterparties; changes in industry norms; and adverse outcomes in legal or other dispute resolution proceedings. If one or more of these risks materialize, or if underlying assumptions prove incorrect, actual results may vary materially from those expected. You should not place undue reliance on forward-looking statements. For a more complete discussion of these and other risk factors, please see the company's most recent filings with the Securities and Exchange Commission, including its annual report on Form 10-K for the year ended December 31, 2017 and subsequent quarterly reports on Form 10-Q. This presentation reflects the views of the company's management as of the date hereof. Except to the extent required by applicable law, the company undertakes no obligation to update or revise any forward-looking statement.

Gasoline Alkylate

- **Excellent Blending Component**
- **Highly sought after in the marketplace**
- **Alkylate Characteristics**
  - Clean combusting iso-paraffins
  - High Octane Value
  - Low Vapor Pressure
  - No Olefins
  - No Aromatics
  - Low Sulfur Content

ALBEMARLE

LUMMUS TECHNOLOGY 3

Gasoline Market Requirements

- **Market Driven:**
  - Global gasoline demand
  - New Engine Standards:
    - Higher contribution of Alkylate to meet increasing octane demand
  - Abundance of "Cheap" Feedstock (USA & Middle East)
- **Environmentally Driven:**
  - Tighter Fuel Specifications & Mandates (Sulfur, Aromatics, etc...)
  - Phasing out of HF Alkylation units for safety reasons

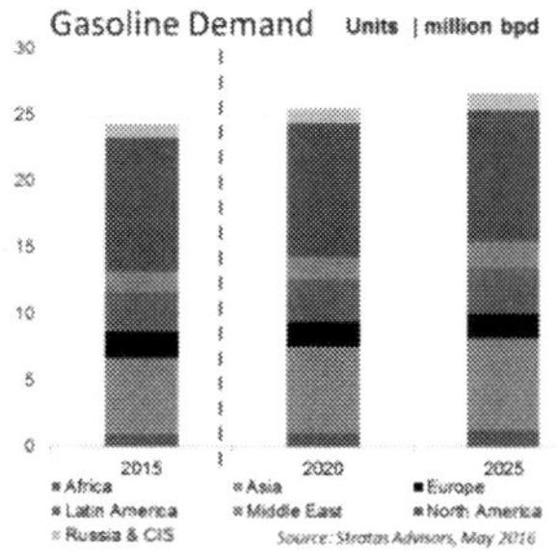

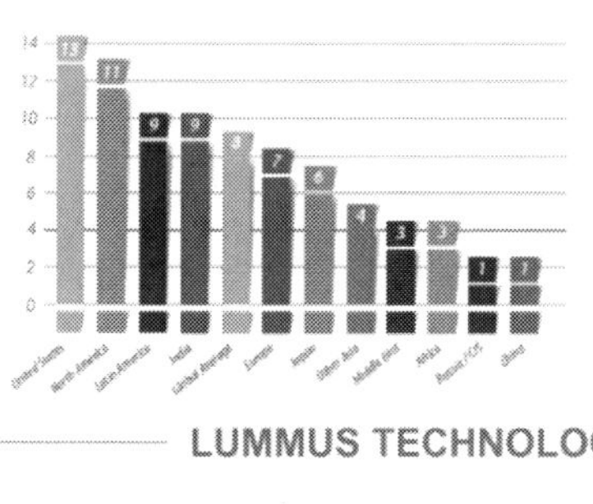

ALBEMARLE

LUMMUS TECHNOLOGY 4

Novelty of AlkyClean Technology

- Completely eliminate the use of liquid acids and associated hazards
- Provides a credible solution for HF Alkylation unit conversion

- **Hydrofluoric acid (HF) is extremely toxic**
  - Lethal clouds forms for miles upon release
  - Incident at Gumi, S. Korea, September 27, 2012:
    - Unloading HF to storage tank
    - 5 people died & 18 injured
    - >3000 people requested medical attention
    - 300 people displaced from homes
    - Agricultural damage & livestock affected

Solid Alkylation Process Challenges and Opportunities

- Process challenges for solid catalyst alkylation technology:
  - Ensure alkylate quality is on par with conventional technologies.
  - Create a solid catalyst system with minimal deactivation.
  - Develop an economically viable catalyst regeneration scheme.
  - Develop a process/catalyst system that add value to refiners.

- Process Opportunities for solid catalyst alkylation technology:
  - Completely Eliminate use of liquid acids
  - Completely Eliminate liquid acid catalyst associated hazards
  - Completely Eliminate traditional technologies operational complexity

➢ **AlkyClean technology successfully addresses all of these challenges, and is commercially proven.**

Mechanism of Iso-Butane Alkylation

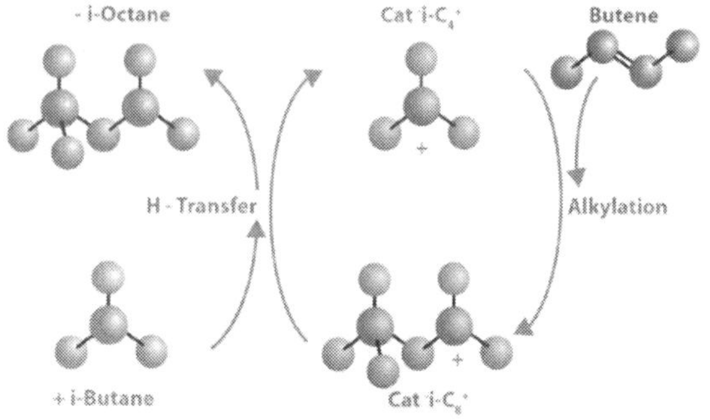

- Formation of C4+ & C8+ ions on the acid sites via Alkylation reaction
- Optimized catalyst pore size and acid sites concentration of our AlkyStar® Catalyst
- Favor the formation of TMP species
- Avoid the formation of heavy species
- Targeted operating conditions that favor the H – transfer pathway and IsoOctane formation

ALBEMARLE LUMMUS TECHNOLOGY 7

A Path to Successful Scale Up…

Proof of Concept at Bench Scale

Demonstration & Validation at Medium Scale

380 MTA

Shandong Wonfull in ZiBo China
(First Licensee Awarded)

100,000 MTA

ALBEMARLE LUMMUS TECHNOLOGY 8

Performance of the Demo Unit

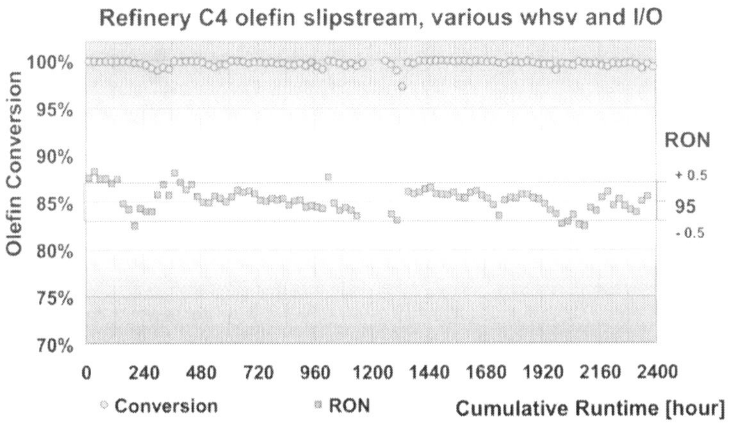

- Stage Approach to Scale-up of our AlkyStar Catalyst & AlkyClean Process

Proven Scale-up of our AlkyStar Catalyst & AlkyClean Process

- AlkyClean Commercial Information

RON	95-97
Sulfur	< 1 ppm
RVP	< 6.5 psi
ASTM D-86 FBP	< 208°C
Overall Conversion of Olefins	99.99

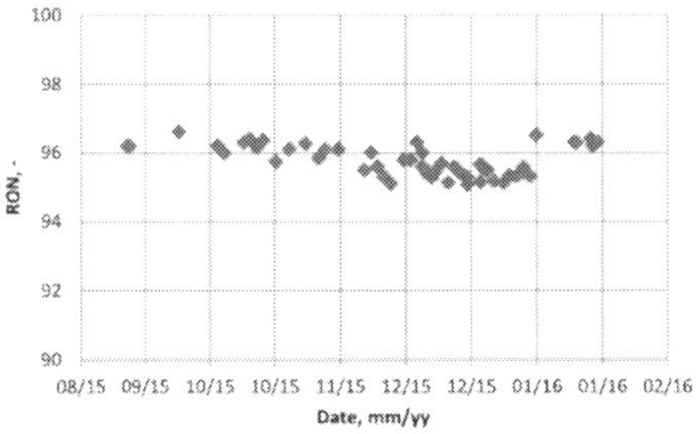

Timeline for Commercialization of AlkyClean First Licensed Unit

- May - September, 2013: PDP Execution
- January, 2014 – June, 2015: Engineering Procurement and Construction
- July, 2015: Pre- commissioning
- August, 2015: Start Up

- Lummus Technology/Albemarle technical support to licensees:
  - All phases of engineering and construction of the Unit.

M ALBEMARLE LUMMUS TECHNOLOGY 11

Commercialization of AlkyClean Technology: The World's First Solid Catalyst Alkylation Unit

- Location: Shandong Wonfull Petrochemical Group Co, PRC
- Capacity: 2,700 BPD of Alkylate production
- Unit Startup: Smoothly and safely started up in August 2015
- Performance test: Completed within 4 months-All guarantees met
- Since Then…The unit has been producing an excellent quality alkylate product

- **Successfully Demonstrated the Scale-up of our AlkyStar Catalyst & AlkyClean Process**

M ALBEMARLE LUMMUS TECHNOLOGY 12

Advantages of AlkyClean Technology

- Technical Advantages: a Breakthrough…
  - No Liquid Acids.
  - Eliminate Corrosion.
  - High Product Quality: High octane & low RVP.
  - Feed Flexible: C3's C4's and C5's olefin streams.
  - Offers flexibility for coproduction of gasoline and kerosene.
  - High Yield due to the Elimination of ASO production.
- Operation Advantages: Simple & Robust Design…
  - Employs simple fixed-bed (FB) reactor system.
  - Well-proven zeolite-based catalyst.
  - Eliminates product post-treatment or washing.
  - Catalyst fully regenerated on a cyclical basis.

ALBEMARLE LUMMUS TECHNOLOGY 13

Advantages of AlkyClean Technology

- Environmental Advantages: A Peace of Mind…
  - No Liquid Acids & Caustic Wash.
  - Eliminates safety & chemical process hazards associated with liquid acid handling.
  - Reduces Environmental Footprint.
- Economical Advantages: Added Value…
  - Reduced Maintenance Costs & Required Turn Arounds.
  - Significantly Reduced Safety Mitigation Costs associated with liquid acids; i.e. HF.
  - Viable option to revamp existing HF Alkylation Units.
  - Economically competitive for Grass Root Alkylation Projects.

ALBEMARLE LUMMUS TECHNOLOGY 14

AlkyClean Process: Simplified Process & Easy Reactor Scale-up

Operating Temperature: 50°C - 90°C and 290-440 psig.

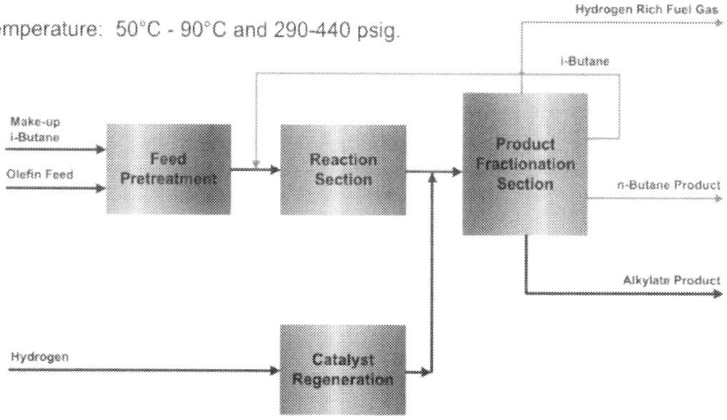

- Employs a simple fixed-bed reactor system with a zeolite-based particulate catalyst
- Multiple fixed bed reactors are used for continuous alkylate production
- Individual reactors cycle between on-line alkylation and low temperature regeneration
- Catalyst is regenerated on a cyclical basis, fully restoring catalyst activity

ALBEMARLE® LUMMUS TECHNOLOGY 15

AlkyClean Process: Successful Scale-up of the Regeneration Scheme

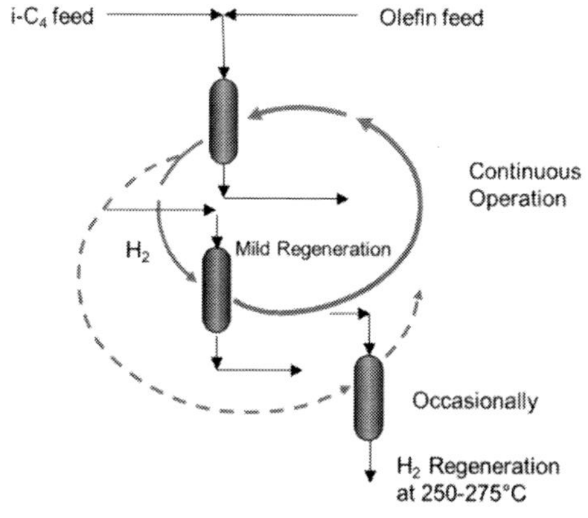

- Multiple reactors are used for continuous alkylate production
- Individual reactors cycle between on-line alkylation, mild and high temperature regeneration
- Catalyst is fully regenerated on a cyclical basis, i.e. restoring catalyst activity

ALBEMARLE® LUMMUS TECHNOLOGY 16

Catalyst of AlkyClean Technology: Albemarle's AlkyStar ®

Albemarle's AlkyStar catalyst **is designed and tailored** for the AlkyClean process:

Albemarle's Catalyst Manufacturing Facility in Amsterdam Where AlkyStar® Is Made

- Uses of well proven zeolite catalyst system
- Low concentration of noble metal
- Optimized strength, type and number of zeolite's acid sites to enhance hydrogen transfer reactions
- Optimized particle size and porosity to overcome any diffusion limitations and heavy species formation

➢ **The Scale-up of our Robust AlkyStar Catalyst is Proven**

AlkyClean Technology Recognition:

Presidential Green Chemistry Award U.S. EPA 2016

44th Kirkpatrick Award 2017

American Chemical Society Award Winner 2010 for Affordable Green Chemistry

ACS Chemistry for Life

Wall Street Journal Technology Innovation Award 2002 – Category Environment

In Summary, AlkyClean Technology .....

- Only proven & commercialized solid catalyst alkylation process in the world
- Produces a high octane, low RVP alkylate on par with conventional technologies when processing C3's C4's and C5's olefin streams.
- Eliminates safety and process hazards associated with liquid acid handling and transportation.
- Inherently safer and environmentally friendly compared to commercially demonstrated alkylation processes.
- Economically competitive for Revamp (HF) & Grass Root Alkylation Projects

➢ **AlkyClean brings Peace of mind to the Refinery Operations**

M ALBEMARLE *Peace of mind* LUMMUS TECHNOLOGY 19

Revamp of Existing HF units:
AlkyClean Technology Solution

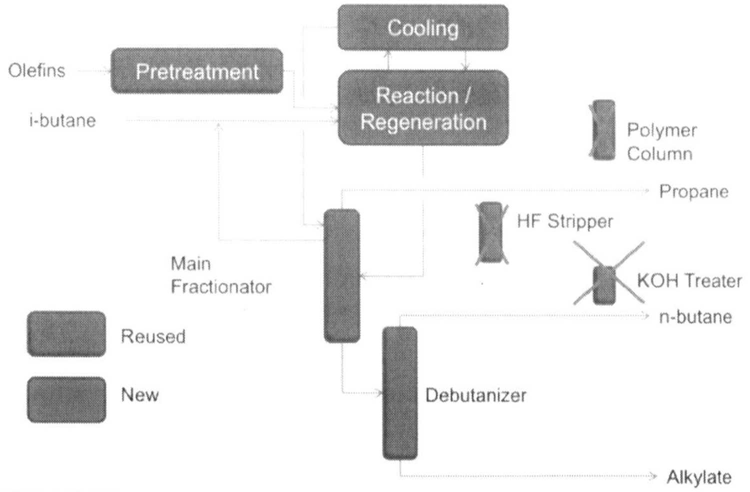

Revamp of Existing HF units:
AlkyClean Technology Incentives

- Continuous cost of HF mitigation: **Eliminated**
- Maintenance Cost: **Significant Reduction**
  - No corrosion issues
  - No replacement of expensive materials
  - Less frequent shutdowns & turn around
- Impact of products on downstream units: **None**
- Operating costs: **Similar**
  - Catalyst instead of HF acid processing
- Product quality: **Better**
  - No ASO = Higher Yield
  - Higher RON, due to insensitivity for Butene-1

Effects of Surfactants on Flotation Separation Performance

Authors:

Shi, F., National Energy Technology Laboratory

Flotation is a continuous, efficient, inexpensive, and well-developed commercial process. It has been widely applied in a variety of industrial processes, including mineral processing, and oily water treatment. Gas bubbles are generated within a solid-liquid or liquid-liquid suspension during flotation process. Based on the hydrophobicity of the surface, hydrophobic particles/immiscible droplets attach easily to gas bubbles and rise to the froth layer on top of the liquid surface. Finally, the floated solid particles, or immiscible liquid droplets, are skimmed off to achieve the separation.

In this study, we have investigated the effects of surfactants on the enrichment of rare earth elements (REEs) from coal and coal by-products, and the cleanup of oily sea water. The effectiveness of surfactants, i.e., frother, depressant, and collector, can strongly influence the flotation process. Different types of frothers and collectors were studied at varying flotation conditions. The properties of particle size, temperature, and pH were also studied. The discussion will focus on the performance of these surfactants and the efficiency of separation at those operating conditions.

Predicting the Effectiveness of Glidants Used to Improve the Flowability of Cohesive Pharmaceutical Powders

Authors:

Capece, M., NJIT

This work focuses on understanding and predicting the flowability of pharmaceutical powders. Because well-flowing powders are critical to the successful manufacturing of tablets and capsules, this topic is of upmost importance in the area of drug product development. Special attention will be given to the role played by flow additives such as nano-sized silicon dioxide which are commonly known as glidants. Fundamental aspects such as inter-particle cohesion will be discussed with an emphasis on the relationship between material properties, particle-scale interactions, and bulk-scale behavior. Examples of success and failure experienced in practical application will be given to exemplify the complexity and difficulty in applying theoretical aspects to predicting powder flow behavior.

Development of a New Anti-Foamer for Food Processing

Authors:

King, S., The Dow Chemical Company

In food processing, uncontrolled foaming media can result in a significant loss in production capacity, due to inefficient mixing or pumping, clogged process lines, overflows which result in spills, and product waste. Mechanical methods of foam management often have limited effectiveness. The addition of foam control agents is considered to be more practical and is more widely employed to minimize production losses due to foaming. A foam control agent can include a defoamer and an antifoamer. An antifoamer is designed to prevent foam, whereas a defoamer eliminates existing foam.

Potatoes have a dry substance content of about 25 % which consists mostly of starch, crude fibers and proteins. During processing of potatoes, foam is generated because foam active substances (starch and protein) are extracted by water during transport, washing, peeling and cutting. Like potatoes, sugar beets have a dry substance content of about 25% and also contain saponin, a bio-surfactant, which can cause foaming during processing. In the case of sugar beet processing, some of the foam generated by saponin can be stabilized by sugars. As a result a foam control agent which is useful for potato processing may not be useful for sugar beet processing.

One class of foam control agents used for food processing include block copolymers composed of ethylene oxide, propylene oxide, and/or butylene oxide. These types of products are believed to be effective since at increased temperatures, they are insoluble in solution, thereby causing an increase in the surface tension of the system, which results in foam collapse. Another class of foam control agents for food applications include polydimethylsiloxane (PDMS) materials. PDMS polymers are effective because they have low surface tensions, and are also highly insoluble in water. However, growing customer demand for solutions based on renewable materials has spurred the development of new more bio-based foam control agents. In this presentation, several new classes of materials which are, at least in part, bio-based and showed excellent antifoaming properties for sugar beet or potato processing will be discussed.

Small Volume Continuous Processing for the Manufacture of Active Pharmaceutical Ingredients

Martin Johnson, Senior Engineering Advisor, Eli Lilly and Company

Scott May, Carla Luciani, Jennifer Groh, Luke Webster, Joel Calvin, Jeff Roberts, Rick Spencer, Ray Boyse, Rob Moylan

2018 AIChE Process Development Symposium, Oak Brook, IL, June 7

New Facility for Fully Continuous Drug Substance Processes

Investment in Lilly Manufacturing, Kinsale Ireland
Facility completed Q3 2017
First project in new facility Q4 2017-Q1 2018

Small Manufacturing Footprint

2018 AIChE Process Development Symposium, June 5-7, Oak Brook, IL

3

Lilly Small-Volume CM Facility

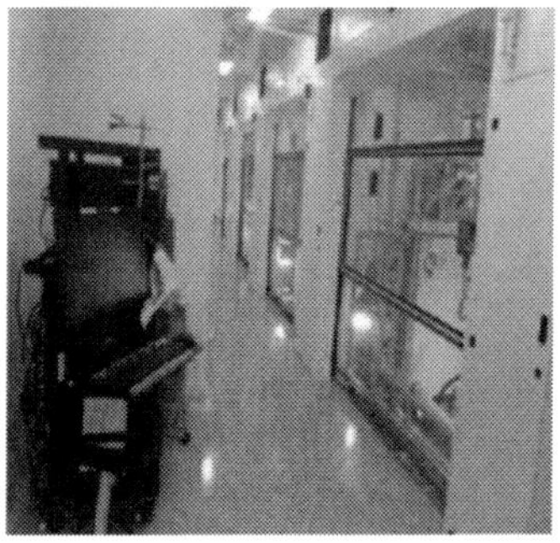

- Skids in fume hoods used for reactions and processing
- Process analytical on the floor and QCL support in near by lab

- Facilities for feed preparation
- Chemical processing in fume hoods
- Finishing suite for isolating API

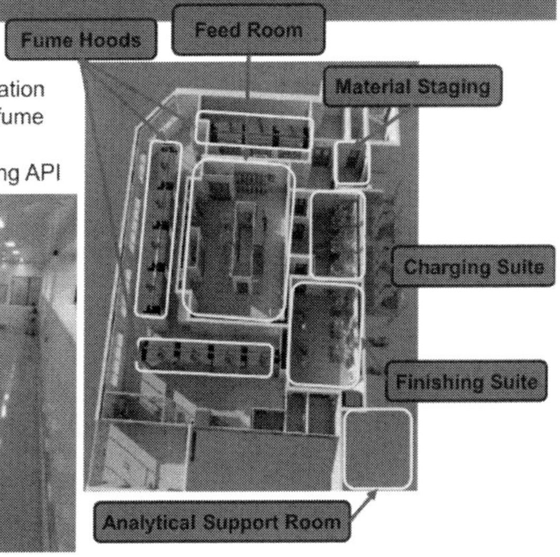

2018 AIChE Process Development Symposium, June 5-7, Oak Brook, IL

Synthetic Route

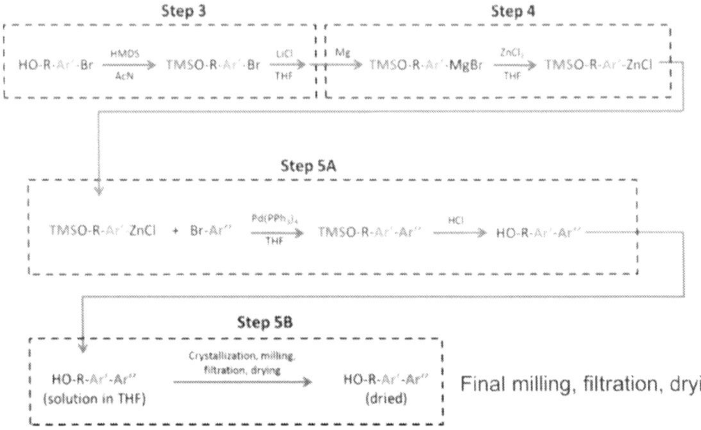

Final milling, filtration, drying is batch

2018 AIChE Process Development Symposium, June 5-7, Oak Brook, IL

5

Continuous Unit Operations & Analytical Controls

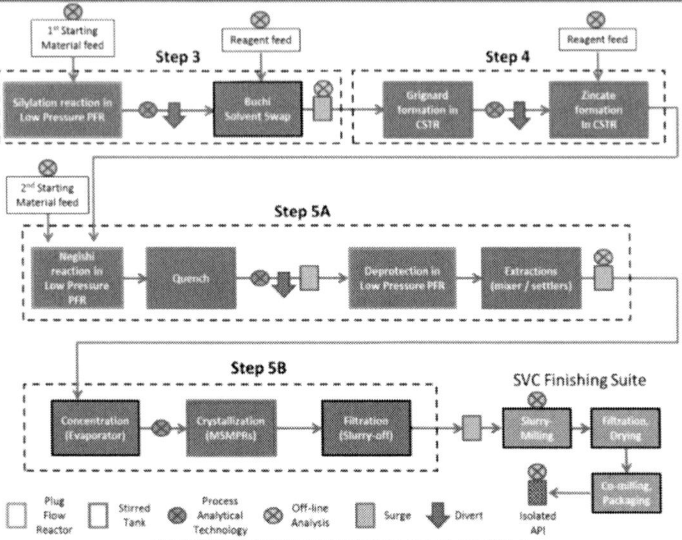

2018 AIChE Process Development Symposium, June 5-7, Oak Brook, IL

6

Continuous Unit Operations & Residence Times

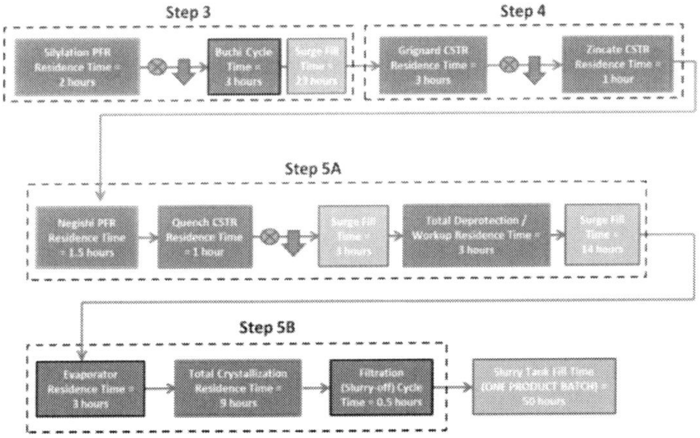

Contrast with drug product CM residence time and RTD.

Continuous Reactors, PFRs

Silylation

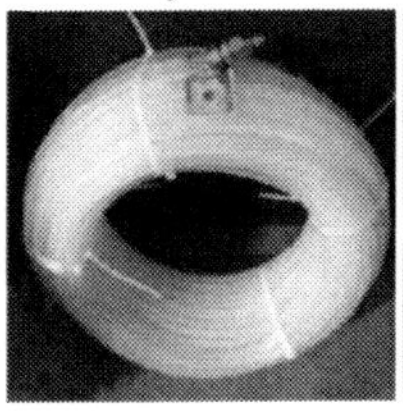

9.9 L PFA tube,
0.75" od 0.625" id
50 m, 120 minute τ

Negishi

24 L PFA tube,
0.5" od 0.376" id
330 m, 75 minute τ

Deprotection

9.6 L PFA tube,
0.5" od 0.376" id
134 m, 28 minute τ

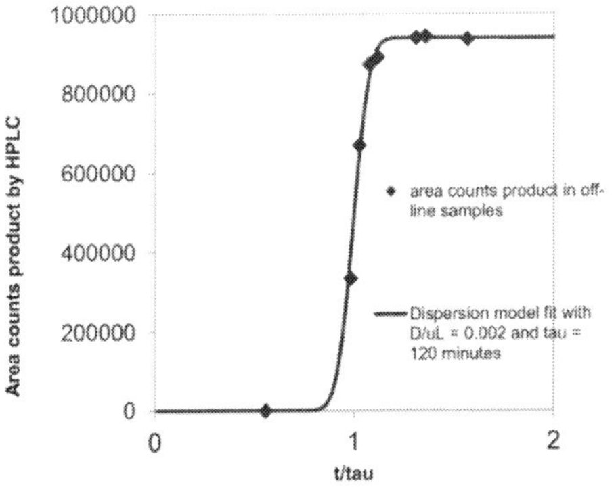

$$\frac{D}{uL}\frac{d^2C_A}{dx^2} - v_x\frac{dC_A}{dx} - kC_A = 0$$

$$\frac{\partial C}{\partial \theta} = \left(\frac{D}{uL}\right)\frac{\partial^2 C}{\partial z^2} - \frac{\partial C}{\partial z}$$

$$C = \frac{1}{2\sqrt{\pi(\frac{D}{uL})}}\exp\left[-\frac{(1-\theta)^2}{4(\frac{D}{uL})}\right]$$

Levenspiel 1962

CSTR Mixer at PFR inlet

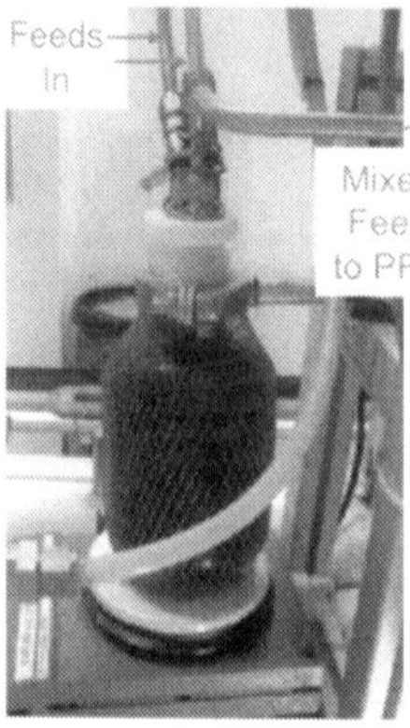

Selection of mixer type and size was important to impurity control strategy.

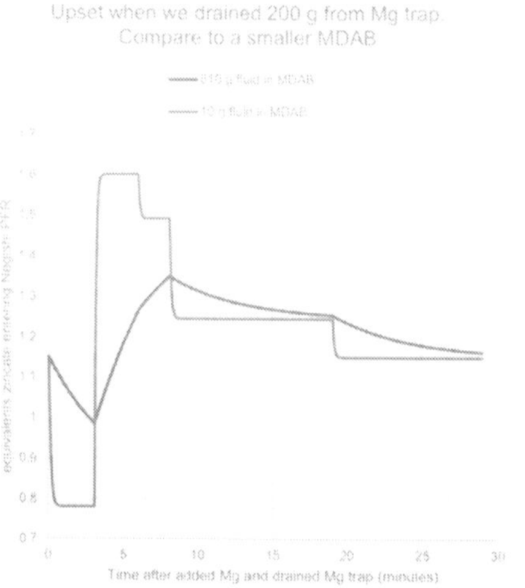

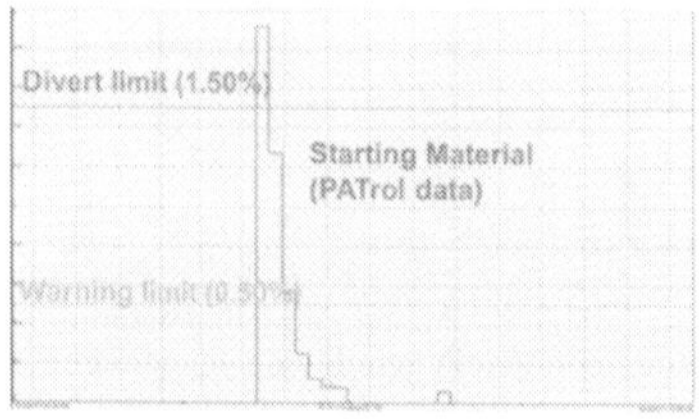

Spike in unreacted aryl Br would have been much worse without the mixer at the PFR inlet.

Continuous Reactors, CSTRs

Girgnard	Zincate	Deprotection

10 L glass 3 hour τ	5 L glass 31 minute τ	10 L glass 31 minute τ

Mg Sequestering

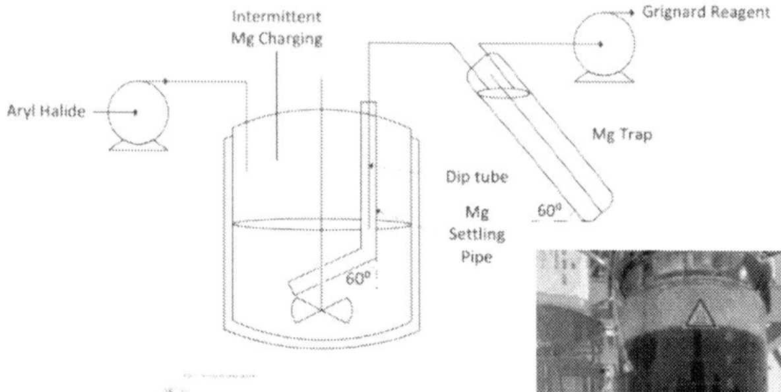

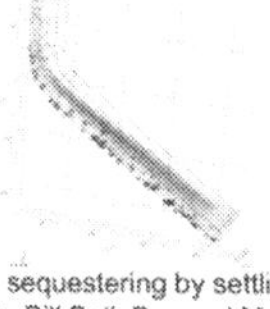

CFD for Mg sequestering by settling pipe.
Rich Shields, Bill Bell, Bernard McGarvey

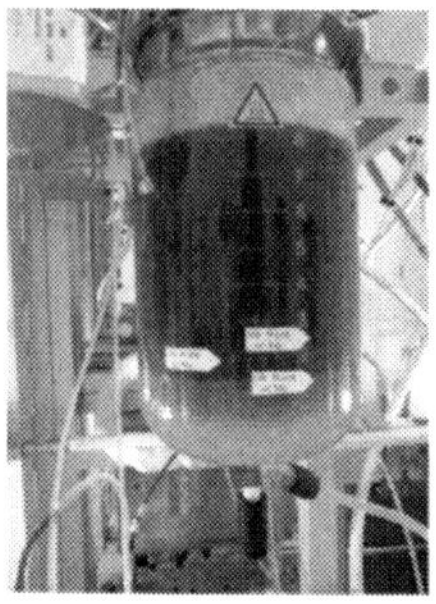

On-line HPLC for each of the main reactions

Silylation

Grignard

Negishi

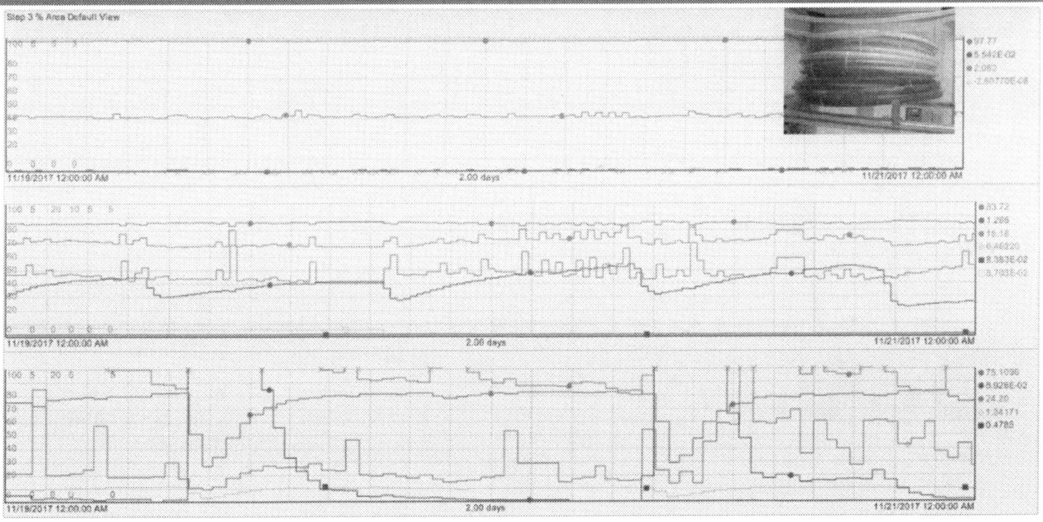

On-line HPLC

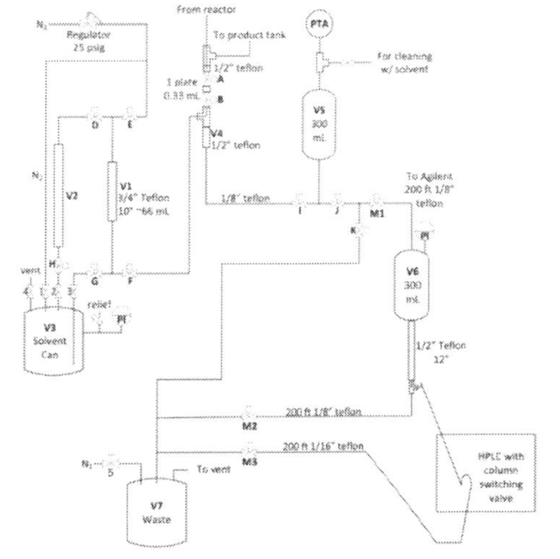

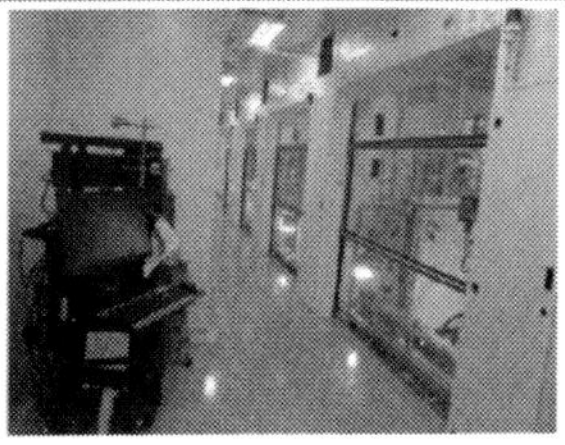

Contrast with PAT analytical frequency of DP CM. 1 hour sufficient frequency for DS CM.

Org. Process Res. Dev., **2016**, 20 (5), pp 888–900, Figure S25

Mixer-settler extractors

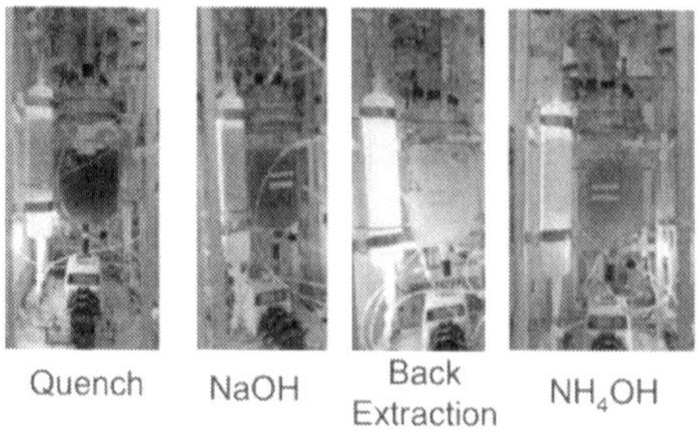

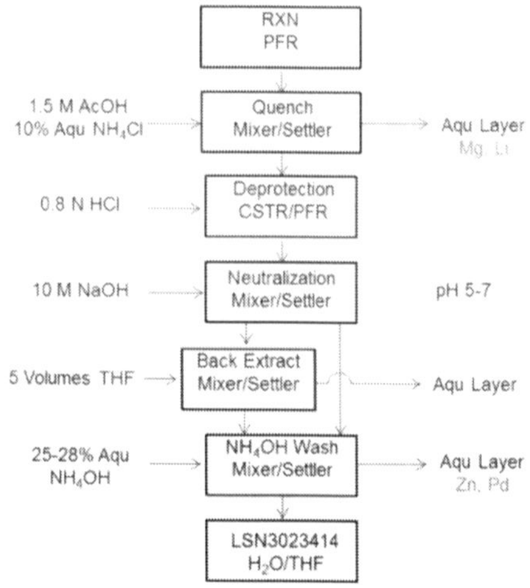

	Zn	Si	Li	Mg	Pd	P
Aqu Quench	7 %	0 %	74 %	93 %	0 %	0 %
Aqu HCl/NaOH	15 %	0 %	21 %	6 %	0 %	0 %
Aqu NH_4OH	77 %	0 %	5 %	1 %	88 %	8 %
Distillate	0 %	94 %	0 %	0 %	0 %	0 %
Isolated solid	0 %	0 %	0 %	0 %	0 %	0 %
Filtrate	1 %	6 %	0 %	0 %	12 %	92 %

Continuous distillations

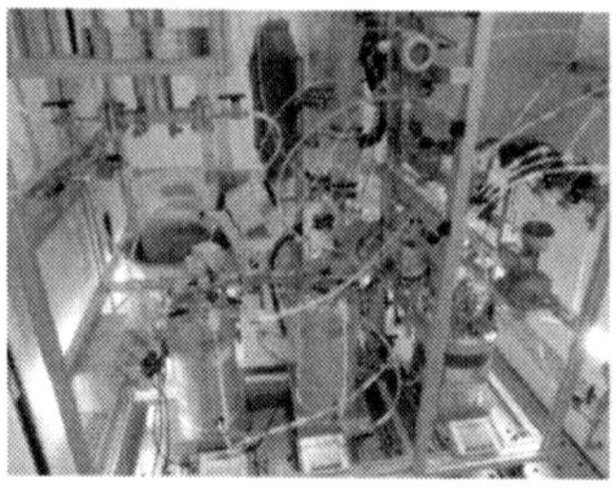

20L rotary evaporator skid.

Strip to dryness.

Remove acetronitrile and HMDS prior to Grignard formation.

Consistent concentration into Grignard.

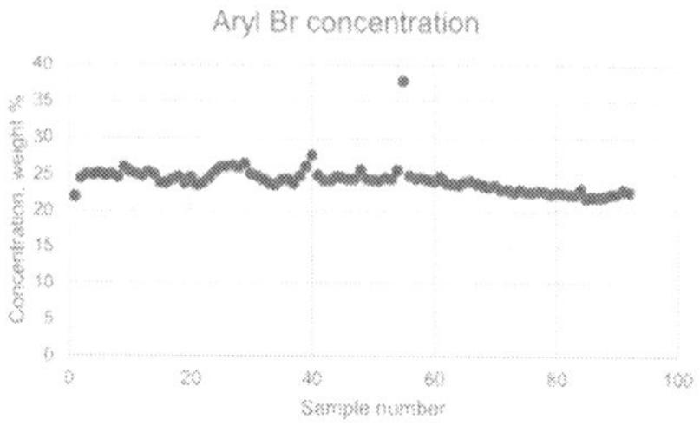

10L CSTR evaporator skid

Consistent API, THF, water concentration going into continuous crystallization.

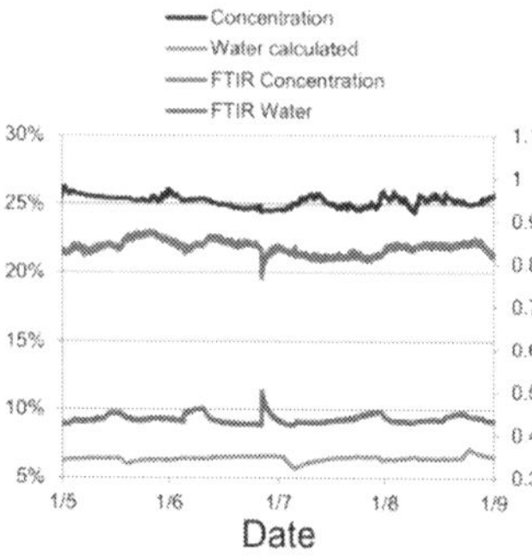

Crystallization

Evaporator	MSMPR 1	MSMPR 2	MSMPR 3	Slurry Off
In at 506%, out at 28% API In at 8-9%, out at 6-6.5% H_2O	60% of API Crystallized 2.7% H_2O	72% of API Crystallized 1.4% H_2O	82% of API Crystallized 1.4% H_2O	Every 30 minutes 165 g of washed API in ACN sent to Surge Collection Tank

Impurity Rejection: Crystallization Performance

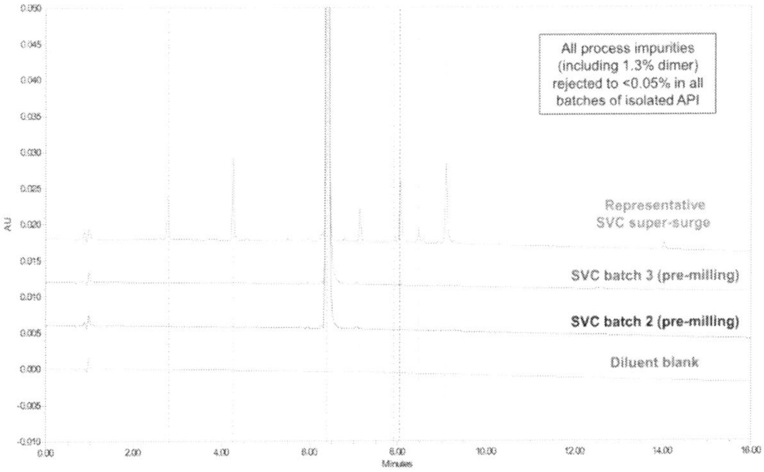

Crystallization Yield

79% yield at steady state after working out the bugs, super surge to end.
Previous campaign batch crystallization was 68.3% (81.45% tech and 83.9% recryst)

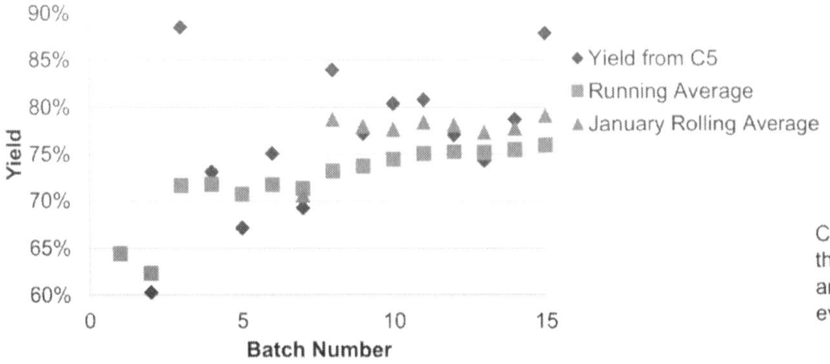

Contrast with DP CM that may have startup and shutdown for every batch.

Process Performance and Robustness

Manufacturing request was 1 week operating time before cleanout.
The process achieved 2 weeks.
Quick cleanout of msmpr1. Short downtime.

Contrast to DP CM. DS ease of stopping and restarting, for example to clean out just MSMPR1 and then keep going.

Surge points

- Decoupling the unit operations.
- Simplifying operational logistics.
- Simplifying startup and shutdown transitions.
- Simplifying automation and control.
- Cleanouts of crystallizers.
- Providing distinct points where the quality of the material can be assessed.
- Dampening out process disturbances
- Facilitating conservative calculations of acceptable magnitude and duration of disturbances.
- Facilitating real time mass balance calculations in the middle of the train.

Contrast with drug product CM which has no surge/decoupling in the middle.

RTD Model Library

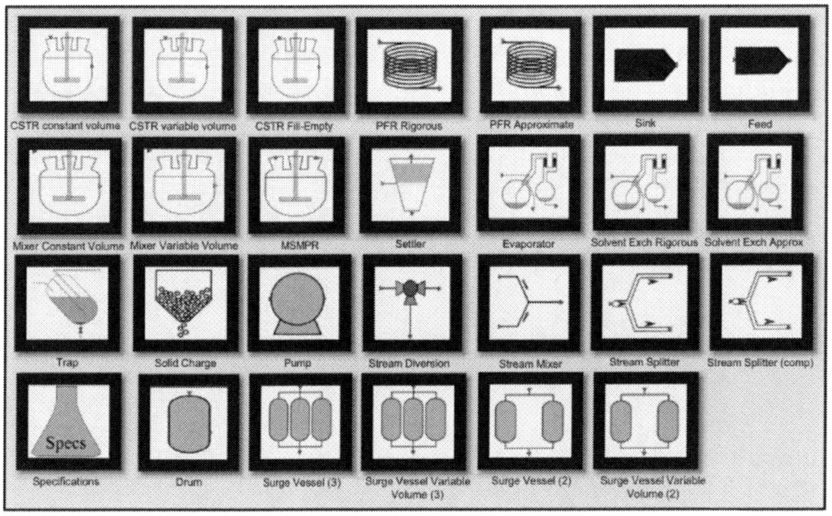

gPROMS

Pi3k Model

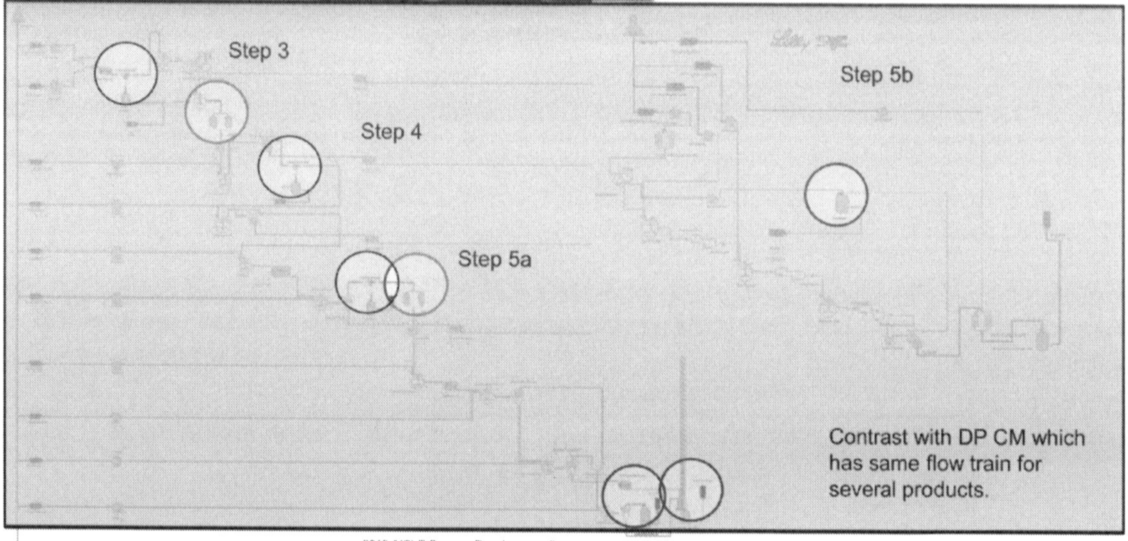

Conclusions

- Flow enabled route because of Grignard unstable intermediate
- Safety benefit of running Grignard flow. Only 10L reactor.
- Small amount of material at risk in reactors at any time.
- Disposable reactors and separators
- Continuous crystallization was better for kinetic impurity rejection.
- 2 weeks msmpr operating time before cleanout
- Process containment in fume hoods.
- Numerical modeling is necessary for RTD, lot genealogy, and control strategy.

Acknowledgements

Continuous Manufacturing for Drug Substance

Small Molecule Design and Development

Martin Johnson, Scott May, Jeff Roberts, Tom Wilson, Carla Luciani, Pete Houghton, Jennifer Groh, Bret Huff, Luke Webster, Paul Collins, Shujauddin Changi, Heather Weimer, Steve Myers, Sarah O'Keeffe, Randy Lambertus, Ahmad Almaya, Jon Dieringer, Candice Wong, Scott Sharp, Todd Maloney, Alison Schmidt, Derek Berglund, Jonas Buser, Eric Moher, Matthew Embry, Joe Martinelli, Richard Miller, Bryan Castle, Bruce Parker, Wei-Ming Sun, Ed Conder, Tim Braden

Lilly Manufacturing – Kinsale, Ireland

John Murphy, Robert Moylan, Paul Desmond, Ray Boyse, Karl Kiely, Regina Lynch, Eleanor Hegarty, Donal Harrold, Noel Sheehan, Stephen Jeffery, Niamh Brody, Nessa Mullane, Brian O'Connell, Kieran Kearney, Colin Cronin, Harry Moloney, James Grace, Edmond Kennedy, Derek Cumiskey, Fintan Ryan, Margaret Harris, Gordon Flanagan, Aoife Corrigan, Diane Tennenhouse, Humphrey Moynihan, Sharon Bourke, Angela Bennett

D&M Continuous Solutions

Jim Stout, Ed DeWeese, Paul Millenbaugh, Doug McKinney, Ed Chow, Morgen Rosemeyer, John Schafer, Jonathan Adler

Opportunities and Challenges for Process Intensification: An Industrial Perspective

Authors:

Calverley, T., The Dow Chemical Company

Despite more than a century of developments in chemical processing, large opportunities for energy and atom efficiency remain. Increasing population means that demand for products of the industry will continue to grow, posing a challenge to shrink the environmental footprint per unit of production.

This talk will examine some of the process improvement opportunities of interest to Dow and how we have attempted to address them. In addition to highlighting some success stories, the talk will present some of the barriers we encounter when attempting to innovate and intensify process technologies.

the Energy to Lead

Application of Aerospace Liquid Propulsion Technology for Process Intensification

Tony Eastland
Director, R&D
Gas Technology Institute

2018 AIChE Process Development Symposium
Oak Brook, IL June 5-7

gti

TEAMWORK

"Teamwork is the ability to work together toward
a common vision. It is the fuel that allows
common people to attain uncommon results."
- Andrew Carnegie

Process Intensification Through High Energy Density Equipment

> Diversity of thought and experience can help solve the challenges in adjacent industries

> Gas Technology Institute (GTI) is applying high pressure combustion technologies acquired from Aerojet Rocketdyne to reduce costs and increase yields/performance in the Chemical, Oil and Gas, and Power industries
  – R-GAS™ Compact Gasifier
  – sCO_2 Power Generation Cycles
  – Direct Contact Steam Generators

3 gti

Aerospace Propulsion Technology Can Enable High Energy Density Components

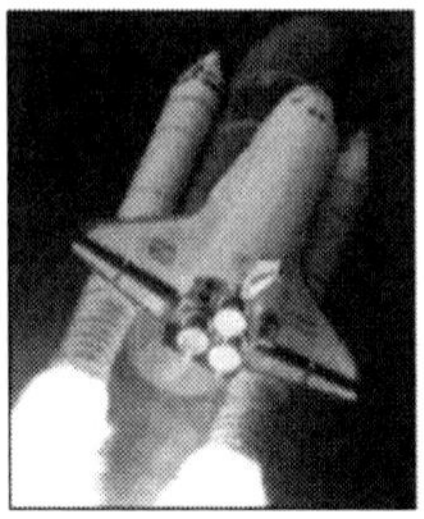

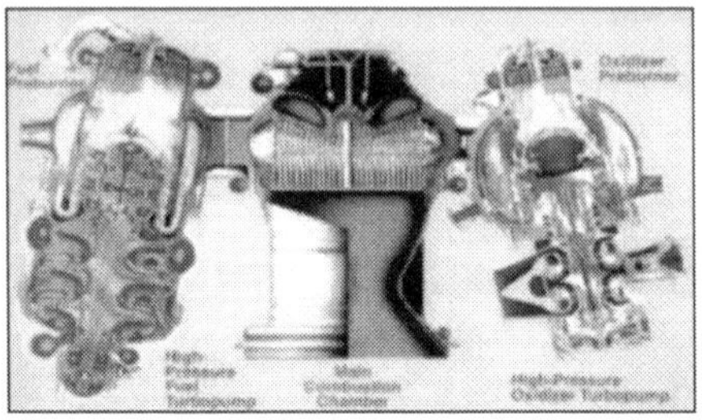

- High energy density combustion
  - 6000°F, 7000psia (3300°C, 48MPa)
- High heat flux cooling
  - Safe metal temperatures
- Short residence time reactions & control
  - Effective mixing

- High speed, high energy density rotating equipment
  - 36,000rpm, 100HP/lb (165kW/kg)
- Low cost < $10 per kW thermal
- Verified design and development processes

4 gti

Rocket Engine Experience Enables R-GAS™ Compact Gasifier

Rocket Engine Experience

- Rapid Mix Injector
- Cooled Membrane Wall
- Rapid Spray Quench

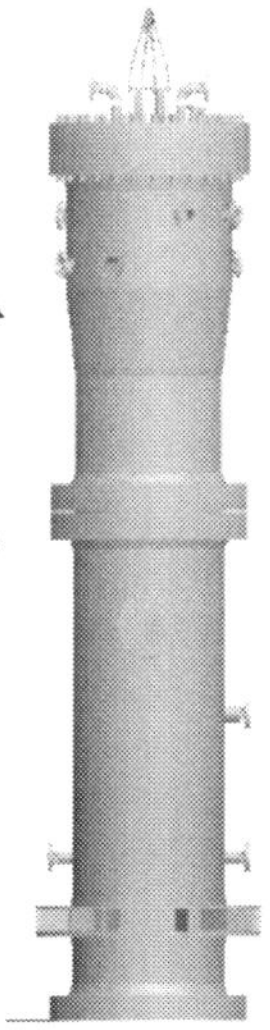

- High flame temperature enhances gasification reactions and enables gasification of high ash fusion temperature (>1600°C) coals without fluxants
- High heat flux cooling system ensures long life gasifier components
- Uniform flow distribution (plug flow) promotes high carbon conversion
- High pressure reduces size

5 gti

R-GAS™ Gasification System

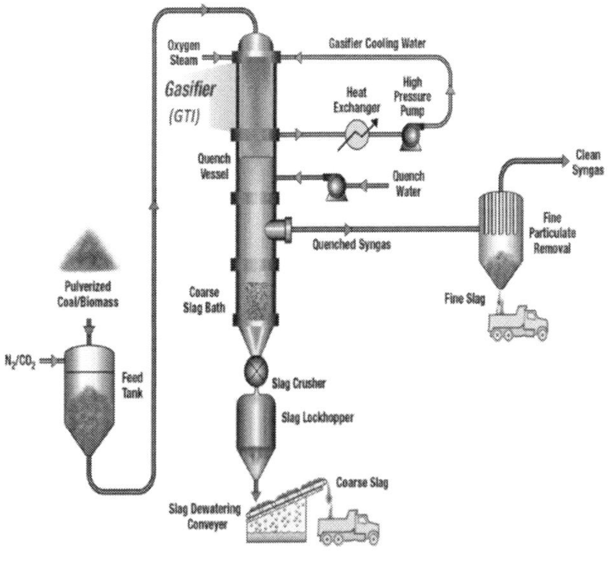

R-GAS™ Attributes:

- 90% size (gasifier volume) reduction
- 20% reduction in plant cost
- 80-85% cold gas efficiency
- 99% carbon conversion
- Able to gasify all ranks of coal, petcoke
- Environmentally friendly waste
- 18TPD Pilot demonstrated
- 800MTPD Demonstration underway

6 gti

Partial Oxidation for Gas-to-Liquids

INNOVATION

- Partial oxidation (POx) reactor part of Turbo-POx technology for converting gas-to-liquids (GTL)
- POx reactor non-catalytically produces hot synthesis gas from steam, natural gas, and oxygen
- Syngas quenched in close-coupled turbo-expander, instead of in conventional waste-heat boiler
- Much of hot synthesis gas sensible heat recovered as electrical power and high-pressure steam, off-setting OPEX

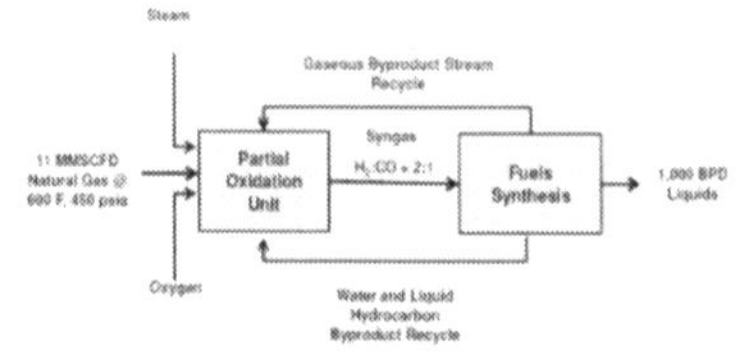

TECHNOLOGY LEVERAGE

- High pressure, high temperature reactions
- Management of high heat flux
- Materials compatibility in high-temperature and high-pressure reducing gas (i.e., syngas) environments under high-stress loading (turbo-expander)

STATUS

- Full-flow POx combustor (and injector) testing at 1,000 barrel/day POGT capacity
- Pilot testing at 400 psia chamber pressure
- Performance mapping at four discrete temperatures and four discrete steam/methane ratios
    - 16 total tests, 4 tests at sooting conditions

7 gti

sCO_2 Cycles Offer Potential for High Electrical Power Generation Efficiency

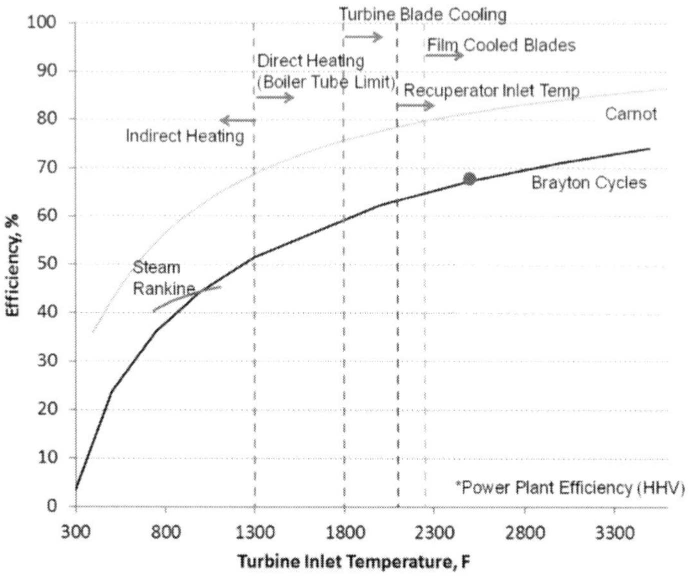

- **High temperature turbine technology at high pressures**
- **Higher cycle efficiency than steam Rankine cycles above ~ 1000°F (540°C)**
- **Efficiencies increase with higher pressure (~ 2-3 points per 1000psi up to 4000psi for both direct and indirect cycles)**

8 gti

Indirectly Heated Recompression Brayton Cycle

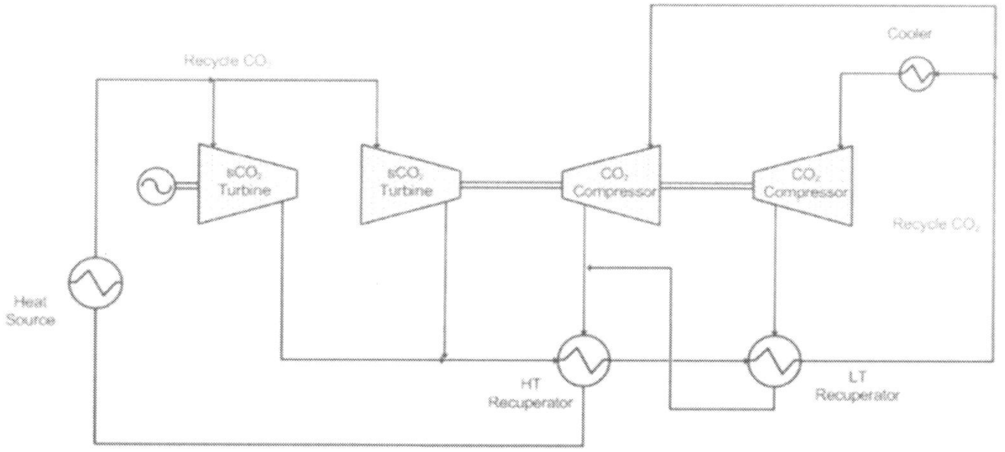

9 gti

Direct Fired sCO_2 Brayton Cycle

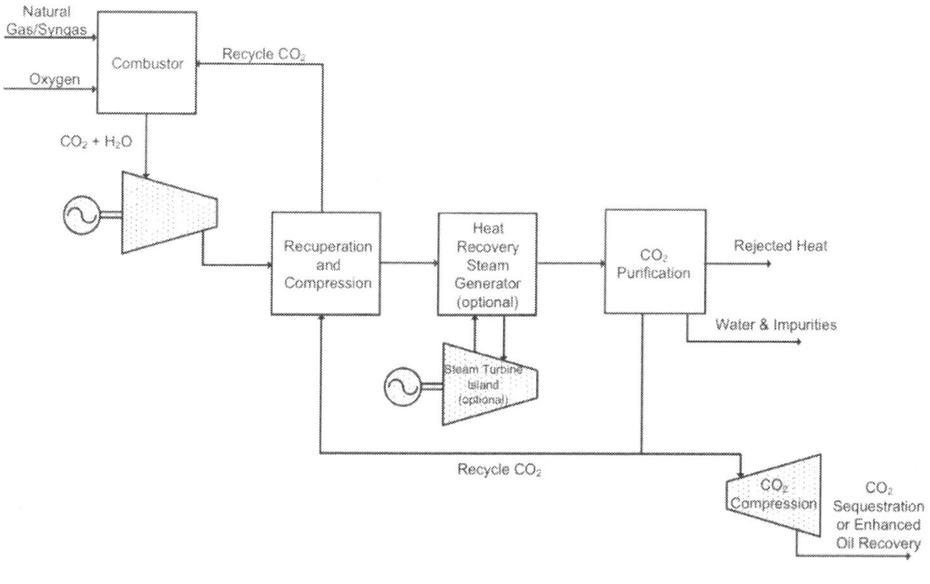

10 gti

sCO_2 Turbomachinery will Combine Steam, Gas, and Propulsion Design Experience

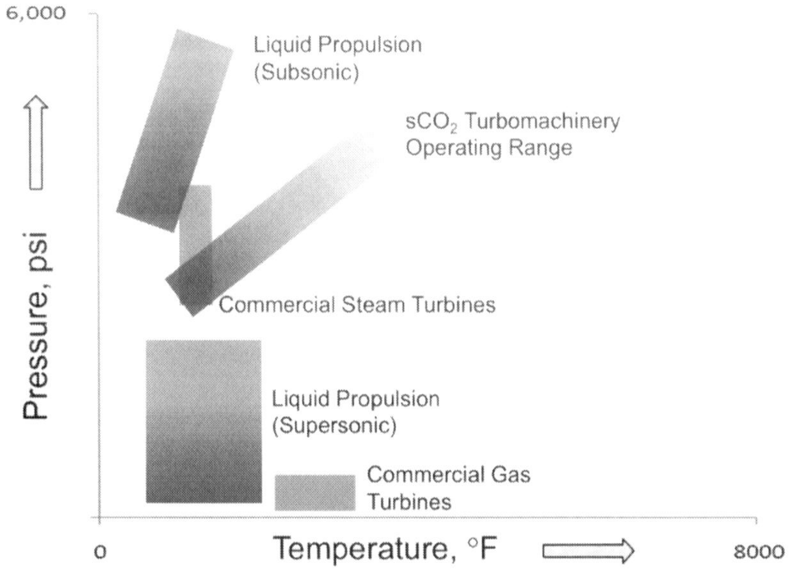

Significant size and cost reduction for sCO_2 vs Steam Rankine cycle turbines

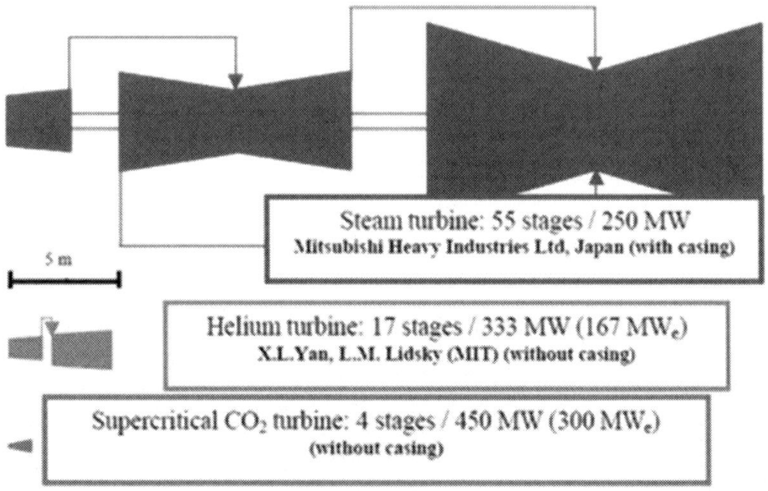

Direct Cycle Combustors will Benefit from Liquid Propulsion Design Experience

> **Direct cycle combustor burns NG or Syngas in O_2 with sCO_2 as coolant**
  - 2000 – 4000psi
  - 1200-1600°C

> **Pressures and temperatures within demonstrated liquid propulsion experience**
  - Up to 3600psi for Methane combustion
  - Up to 7000psi for other fuels (Hydrogen, Kerosene)

> **Several differences being addressed by current research**
  - High Oxidizer inlet temperature (mixed temp of ~750°C)
  - Reaction kinetics at high pressure with CO_2 diluent/coolant

High Pressure Combustion for Direct Contact Steam Generation

INNOVATION

- Waste water treatment with co-production of steam/power
- Compact downhole steam generator using methane combustion for oil sands production
- Compact surface steam generator with lime slurry injection for post-combustion carbon capture

Surface Steam Generator

Downhole Steam Generator

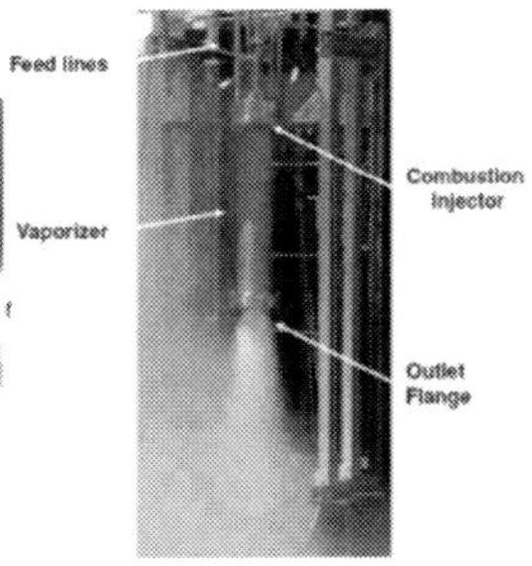

ADVANTAGE

- Compact, robust design
- ~50% CAPEX savings - elimination of water treatment
- >10% reduction in natural gas usage
- ~10% reduction in water consumption

Surface
- Able to use produced water with reduced treatment levels
- Can separate CO_2 prior to injecting into a reservoir
- ~1000psia

Downhole
- Eliminate heat losses by placement at reservoir
- Less water treatment required
- 1500 – 3500psia

STATUS

- Markets: deep/heavy enhanced oil recovery, surface steam injection with CO_2 capture (for oils sands, etc.), waste water treatment with co-production of steam/power
- In-ground test of DHSG conducted;
- Pilot scale SSG demonstrated with produced water
- Viability of CO_2 separation demonstrated
- High pressure ignition demonstrated

Multiple Steam Generation Tests Completed in Recent Years

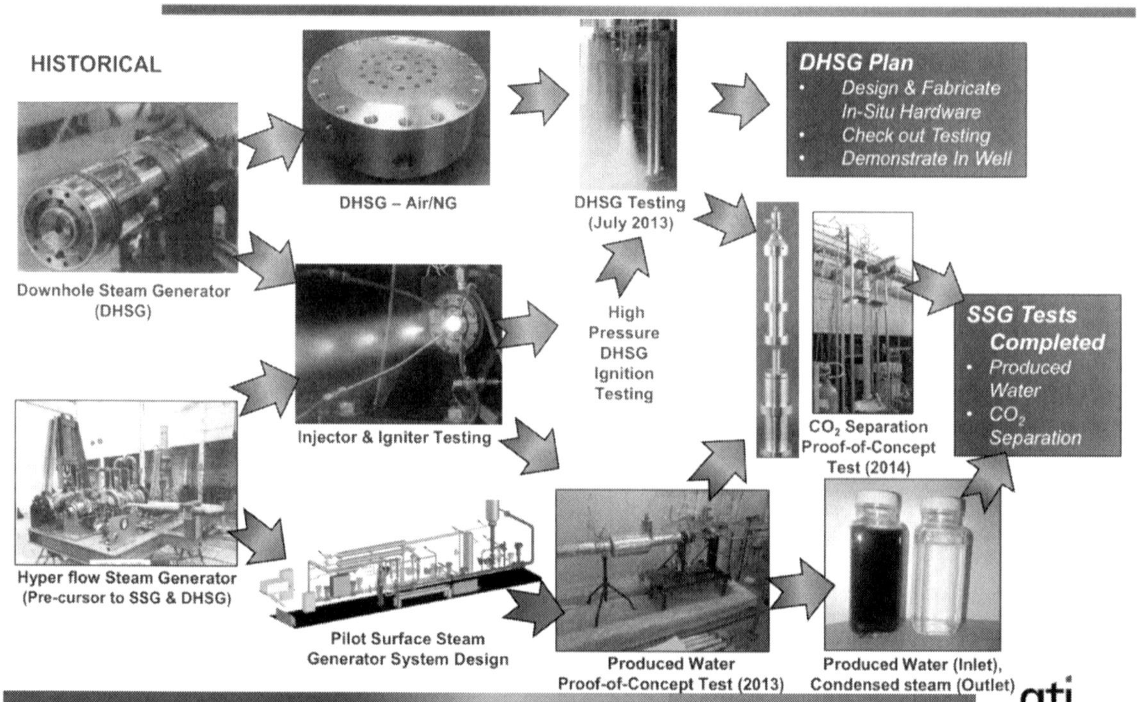

Summary

>Technology development in other industries has often already solved the challenges you face today

>Aerospace liquid propulsion technology is being effectively used to improve performance and reduce costs in Chemical, Oil and Gas, and Power Generation industries through process intensification

TEAMWORK
"Teamwork is the ability to work together toward a common vision. It is the fuel that allows common people to attain uncommon results."
- Andrew Carnegie

Koch Modular
PROCESS SYSTEMS

MODULAR CONSTRUCTION:
An Economic Solution to Regionalized Small Scale Chemical Production

Thomas Schafer

AIChE Process Development Symposium
Oak Brook, Illinois
June 2018

45 Eisenhower Drive, Suite 350, Paramus, NJ 07652 | Tel: (201) 267-8670 | Fax: (201) 368-8989 | www.kochmodular.com

Introduction

Koch Modular
PROCESS SYSTEMS

MODULAR PROCESS SYSTEMS | DISTILLATION | LIQUID EXTRACTION | ENGINEERING | PILOT TESTING

- Economies of Scale Traditionally Have Driven Mega-Scale Plant Economics
- Historically Located in Gulf Coast Because of Raw Material Accessibility
- In Strategic Locations Modular Plant Construction Can Disrupt that Paradigm

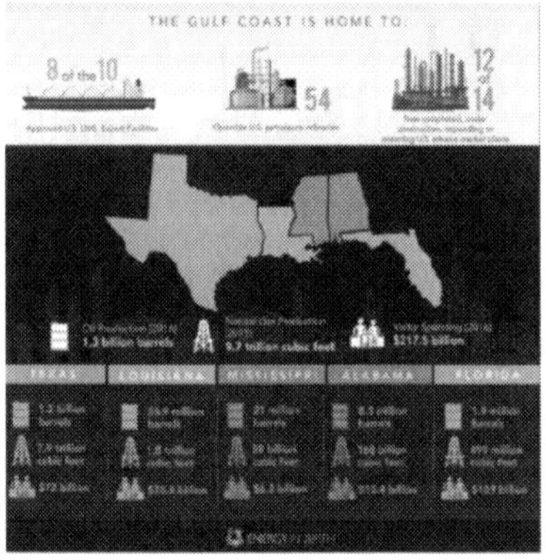

2

www.kochmodular.com
Telephone: (201) 267-8670

Module Assembly

Indoors in Horizontal Orientation

Koch Modular PROCESS SYSTEMS

MODULAR PROCESS SYSTEMS | DISTILLATION | LIQUID EXTRACTION | ENGINEERING | PILOT TESTING

3

www.kochmodular.com
Telephone: (201) 267-8670

Examples of Modular Process Plants

Koch Modular PROCESS SYSTEMS

MODULAR PROCESS SYSTEMS | DISTILLATION | LIQUID EXTRACTION | ENGINEERING | PILOT TESTING

4

www.kochmodular.com
Telephone: (201) 267-8670

Benefits of Modular Construction

Koch Modular
PROCESS SYSTEMS
MODULAR PROCESS SYSTEMS | DISTILLATION | LIQUID EXTRACTION | ENGINEERING | PILOT TESTING

- Shorter Time to Market: Months, not Years, from PO to delivery of modular process units
- Lower Construction Cost: Typically 25-30% lower cost vs. field construction
- Construction proceeds offsite while waiting for permits
- Lump Sum Fixed Price Contracts
- Minimize hazards and safety risks: Work in outdoor conditions and work at elevation eliminated. No hazardous chemicals present
- Single Source Responsibility, Complete System with Mechanical and Process Performance Warranties.

5

www.kochmodular.com
Telephone: (201) 267-8670

Breaking Traditional Economies of Scale Paradigms

Koch Modular
PROCESS SYSTEMS
MODULAR PROCESS SYSTEMS | DISTILLATION | LIQUID EXTRACTION | ENGINEERING | PILOT TESTING

Small Scale Regionalized (non Gulf-Coast) Chemical Production *VS.*
Mega Scale Gulf Coast US Production takes advantage of:

- Low cost of regional raw materials such as natural gas
- Disrupts transportation costs of: 1. Raw materials to Gulf Coast and 2. Product from the Gulf Coast back to regional customers
- Small Scale Designs lead to lower capital expenditure, e.g. simplified heat integration
- Even more so for high pressure systems. Traditional economies of scale less pronounced, reducing "penalty" for not designing at mega scale
- Smaller equipment sizes suited to modular construction and its associated benefits
- Critical surety of supply in potentially volatile market

6

www.kochmodular.com
Telephone: (201) 267-8670

Introducing Two Business Cases

Koch Modular PROCESS SYSTEMS

MODULAR PROCESS SYSTEMS | DISTILLATION | LIQUID EXTRACTION | ENGINEERING | PILOT TESTING

PRIMUS GREEN ENERGY	Petron Scientech Inc.
Small Scale Natural Gas → Methanol	Small Scale Ethanol → Ethylene Oxide
Global leader providing GTL technology through modular systems or licensing	Global technology licensor for conventional and renewable EO and MEG
Solved: Use local natural gas and lower the CapEx Investment Threshold	Solved: Surety of Supply Due to Safety of Transportation Issues

7

www.kochmodular.com
Telephone: (201) 267-8670

Case #1: Natural Gas to Methanol

Koch Modular PROCESS SYSTEMS

MODULAR PROCESS SYSTEMS | DISTILLATION | LIQUID EXTRACTION | ENGINEERING | PILOT TESTING

- This small scale plant design disrupts traditional transportation costs and uses less expensive, locally sourced (Marcellus) natural gas.
- Gulf Coast Transport Model applies in either direction
- Traditional Economies of Scale not as applicable for high pressure equipment
- 20 months to mechanical completion from start of Detailed Design*

*Feed Package Complete

Five Considerations

1. Natural Gas to Methanol
2. MeOH Transportation
3. Economies of Scale
4. Investment Threshold Lowered: CAPEX and OPEX
5. Commercially Proven Technology

PRIMUS GREEN ENERGY

8

www.kochmodular.com
Telephone: (201) 267-8670

Natural Gas to Methanol

Koch Modular PROCESS SYSTEMS

MODULAR PROCESS SYSTEMS | DISTILLATION | LIQUID EXTRACTION | ENGINEERING | PILOT TESTING

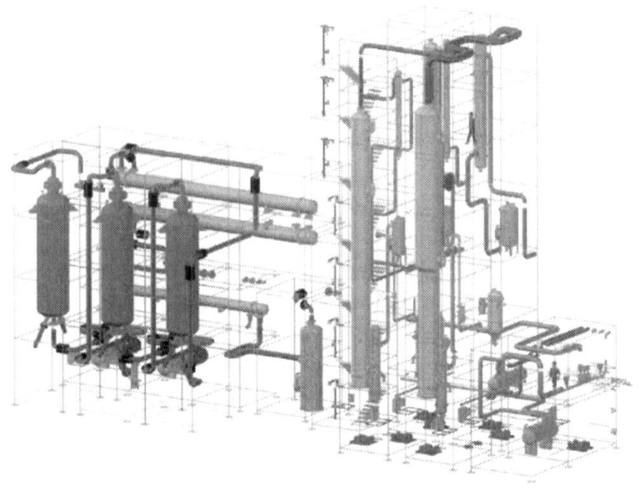

9

www.kochmodular.com
Telephone: (201) 267-8670

Methanol Metrics

Koch Modular PROCESS SYSTEMS

MODULAR PROCESS SYSTEMS | DISTILLATION | LIQUID EXTRACTION | ENGINEERING | PILOT TESTING

Our scale: 100-300 TPD

World/Gulf Coast (Mega) Scale: 2000-5000 TPD

Our Reference Plant Size: 160 TPD

CAPEX: $65MM
(Lump sum total installed cost for modules and balance of plant, leveraging a modest amount of existing site infrastructure)

OPEX: Non-Gas Production Cost: $0.25/gal
Production cost includes labor, electricity, other utilities, routine and major maintenance, catalyst, industrial park site fees, quality control, property tax, insurance, G&A, etc. Excludes natural gas for feed and reformer fuel.

Natural Gas Cost (for feed and fuel): $0.11-$0.33/gal
for $1-$3/MMBTU natural gas price

Schedule: 14 months modules ready to ship, 18 months mechanical completion
for overall site (starting point is detailed engineering/fabrication with the FEED and HAZOP completed)

PRIMUS GREEN ENERGY

10

www.kochmodular.com
Telephone: (201) 267-8670

Case #2: Ethanol—>Ethylene Oxide

Koch Modular PROCESS SYSTEMS

MODULAR PROCESS SYSTEMS | DISTILLATION | LIQUID EXTRACTION | ENGINEERING | PILOT TESTING

- Ensures: Reliable and **safe** supply of Ethylene Oxide
- Gulf Coast mega plant model uses natural gas vs. this model: use of ***readily available ethanol***
- 26 months to mechanical completion*

Five Considerations

1. Ethanol—>Ethylene Oxide
2. Surety of Safe Supply
3. Economies of Scale
4. Investment Threshold Lowered: CAPEX and OPEX
5. Commercially Proven Technology

*Feed package complete

Petron Scientech Inc.

11

www.kochmodular.com
Telephone: (201) 267-8670

Ethanol to Ethylene Oxide

Koch Modular PROCESS SYSTEMS

MODULAR PROCESS SYSTEMS | DISTILLATION | LIQUID EXTRACTION | ENGINEERING | PILOT TESTING

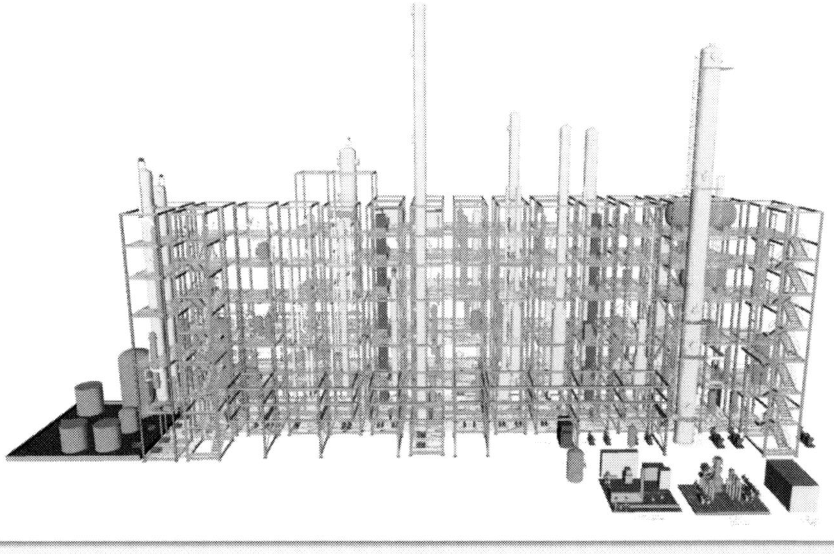

12

www.kochmodular.com
Telephone: (201) 267-8670

Ethylene Oxide (EO) Metrics

Koch Modular PROCESS SYSTEMS

MODULAR PROCESS SYSTEMS | DISTILLATION | LIQUID EXTRACTION | ENGINEERING | PILOT TESTING

Our scale: 10-30 KTA (kilotons per annum)

World/Gulf Coast (Mega) Scale: 100-800 KTA (kilotons per annum)

Our Reference Plant Size: 25 KTA (kilotons per annum)

CAPEX: $120MM
(Lump sum total installed cost for modules and balance of plant, leveraging a modest amount of existing site infrastructure)

OPEX (non-ethanol): $0.09/lb

OPEX (ethanol): $0.29/lb (based upon $1.45/gal plus freight)
Total OPEX: $0.38/lb

Schedule: 22 months modules ready to ship, 26 months mechanical completion
for overall site (starting point is detailed engineering/fabrication with the FEED and HAZOP completed)

Petron Scientech Inc.

13

www.kochmodular.com
Telephone: (201) 267-8670

Koch Modular Process Systems

Koch Modular PROCESS SYSTEMS

MODULAR PROCESS SYSTEMS | DISTILLATION | LIQUID EXTRACTION | ENGINEERING | PILOT TESTING

- Privately held JV company with Koch Industries
- 35 years experience with the process design, detailed engineering and supply of modular constructed process plants
- Renowned for our Expertise in Distillation, Liquid-Liquid Extraction, Process Engineering and Pilot Testing
- Process Conceptualization to Full Scale Manufacturing
- Fortune 500 Chemical Process Industry Client List
- Growing Number of Partnerships with Developing Technology Companies (DTC)

14

www.kochmodular.com
Telephone: (201) 267-8670

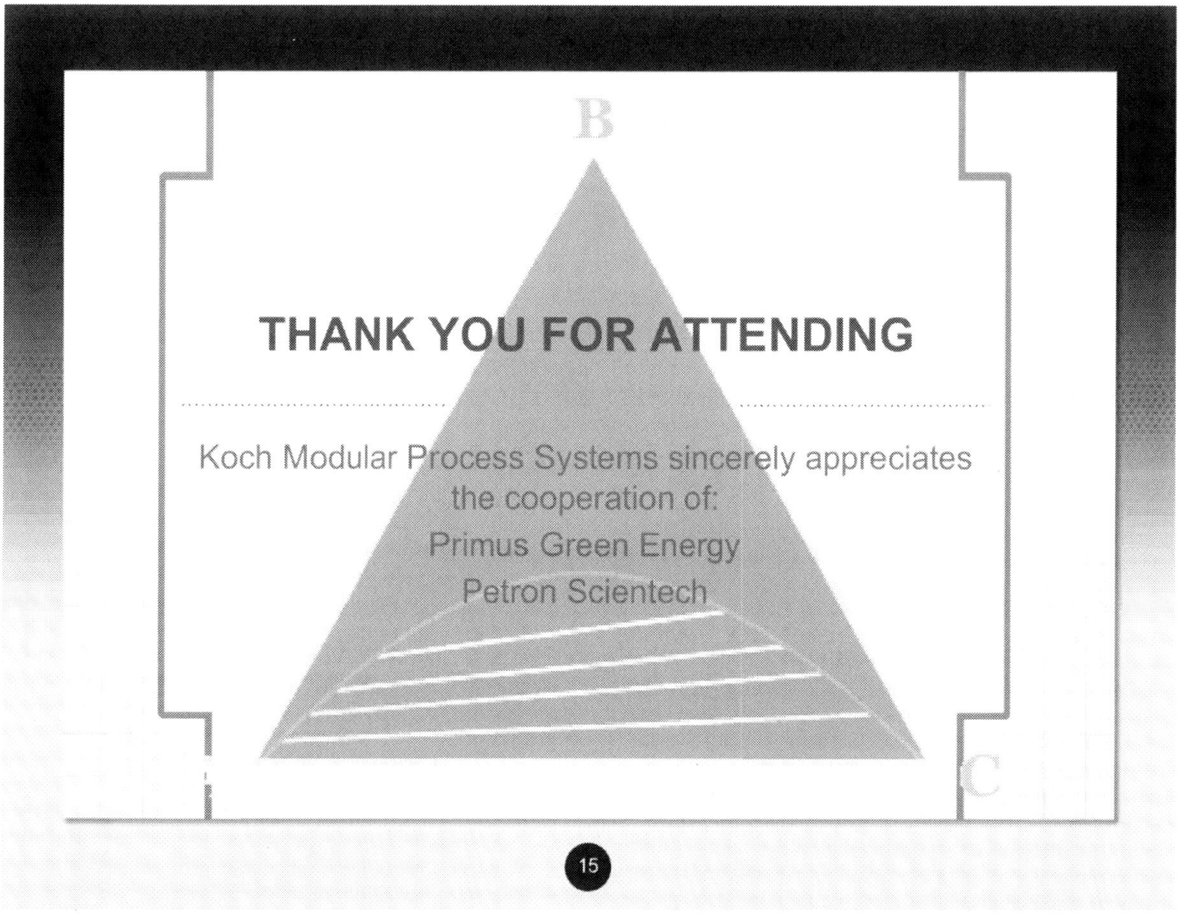
B
THANK YOU FOR ATTENDING
Koch Modular Process Systems sincerely appreciates the cooperation of:
Primus Green Energy
Petron Scientech
C
15

ZETON®

PDS PROCESS DEVELOPMENT SYMPOSIUM 2018

Design Considerations for Deciding Whether to Build Modularly or In-place

Adam Whalley, P.Eng., MBA
Director, Business Development

REALIZE THE FUTURE

Background and Experience

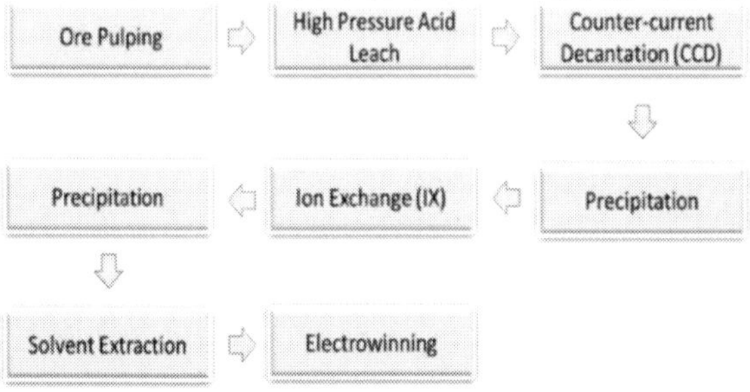

ZETON®

PDS PROCESS DEVELOPMENT SYMPOSIUM 2018

Zeton's Unique Perspective

- Co-located design and fabrication teams
- Unique project execution
- IP protection

ZETON®

pds

Design Considerations for Modular Construction

- Appropriateness of scale
- Inherent process and business characteristics
- Economic and safety risks

ZETON®

pds

Size of Major Equipment

1000 BPD Biodiesel Plant

5000 BPD Production Plant

20 BPD FT Demo Plant

ZETON

pds

Weight of Major Equipment

- Inco demonstration plant
  - Autoclave module exceeded crane capacity
  - Required hydraulic lift
  - Special trailer
- Unrealistic example: $^1/_{20}{}^{th}$ scale hydrocracker:
  - High pressure/high temperature
  - Big diameter, thick heavy walls

ZETON

pds

Transportation Limits

ZETON

pds

Beauty of Pilot Scale

Characteristics	EPCM Approach	Zeton Approach
Communication	Satisfactory	Excellent
Flexibility to Changes	Low	High
Expedited Schedule	No	Yes
Trades Expertise	General	Specialized

- EPCM model works for traditional large scale projects
- Zeton's model works best for pilot, demonstration and some production scales

ZETON

pds

Inherent Process Characteristics

1 BPD Bioethanol Plant

1 BPD DCR Fluidized Catalytic Cracker Pilot Plant

ZETON®

pds

Comparison of Different Technologies

10 BPD FT Demonstration Plant

20 BPD FT Demonstration Plant

ZETON®

pds

Size of the Market/Plant Capacity

3-D Layout - 500 BBL/Day FT Demonstration Plant
- Gasoline market size in N.A. in 2016: $28 billion

Tertiary Butyl Arsene/Tertiary Butyl Phosphene Production Plant
- Global electronic chemicals market size in 2016: $21 billion

ZETON®

pds

Safety of Off-Site Fabrication

OCM Demonstration Plant for Siluria

ZETON®

pds

Drivers for Point of Use

ZETON®

pds

Point of Use Example: Solvay

Smart Scale Satellite Production Unit:
- On-site electronic grade F_2 gas production
- Modularization improved safety

ZETON®

pds

Capital Expenditure – Not a good reason to reduce plant size

- Lower with a smaller scale modular plant
- Reduces amount of capital at risk
- However, risk increases at smaller scale due to economics
- Remember EV = [P(x) * n]

ZETON

pds

Commodities Historical Prices

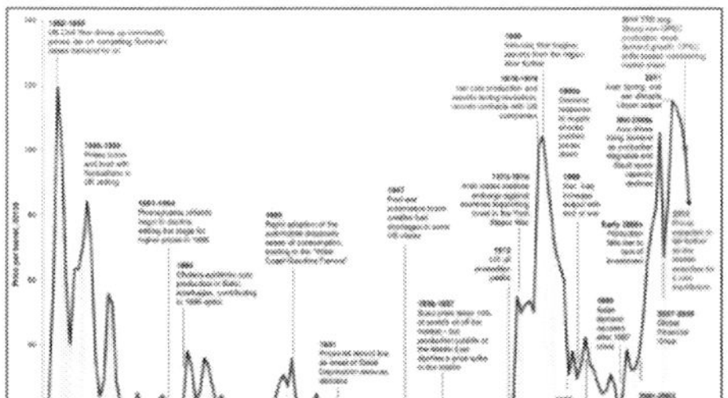

Inflation Adjusted Oil Prices
1861 - 2014

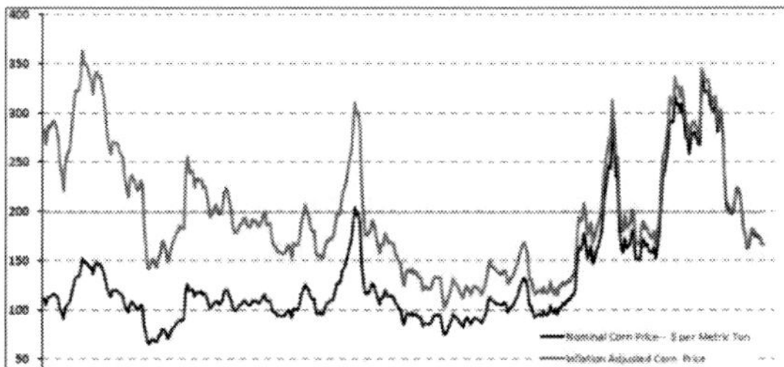

Inflation Adjusted Corn Prices
1981 - 2015

ZETON

pds

Economies of Scale

Pipe	○	○
Diameter	4 in	2 in
Flow Rate	40 L/min	10 L/min
Product Price	1 $/L	1 $/L
Revenue	$40/min	$10/min

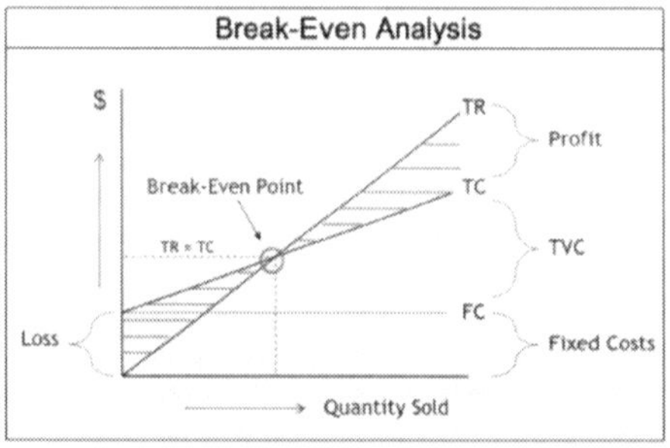

Remember EV = [P(x) * n]

ZETON

pds

Summary Slide

- Appropriateness of scale determined by:
  - Equipment size; both dimensions and weight
  - Transportation issues for the size of the modules
  - Lab, pilot and most demo scales are perfect for modular fabrication
- Inherent business and process characteristics:
  - Technology; market size; remote vs developed infrastructure
- Safety and economic risks:
  - Transportation; ROI; "hot work at site"

ZETON

pds

Questions

Adam Whalley, P. Eng., MBA
Director, Business Development
awhalley@zeton.com
905-632-3123 X235

ZETON®

pds

Safe Chemistry

ZETON®

pds

Considerations for Modular Lab, Pilot, and Demo Scale Plants

- Continuation of on-site operations
  - Avoidance of on-site "Hot work"
  - R&D technology development can proceed in parallel to design and fabrication
- Parties stay within their respective areas of expertise

ZETON®

pds

Two New Modular Systems for GTL(Gas to Liquids) by GTI

- Presented by Terry Marker, Senior Institute Engineer, Bioenergy Initiatives Manager, Gas Technology Institute
- Presented at 2018 AICHE Process Development Symposium
- Oak Brook, IL
- June 7, 2018

gti

GTI Company Overview

- Independent, not-for-profit established by the natural gas industry in 1941
- GTI tackles tough energy challenges turning raw technology into practical solutions
  - Expanding the supply of clean, abundant, and affordable natural gas
  - Transforming all energy resources into clean fuels, power, and chemicals
  - Ensuring a safe and reliable energy delivery infrastructure
  - Promoting the clean and efficient use of energy resources

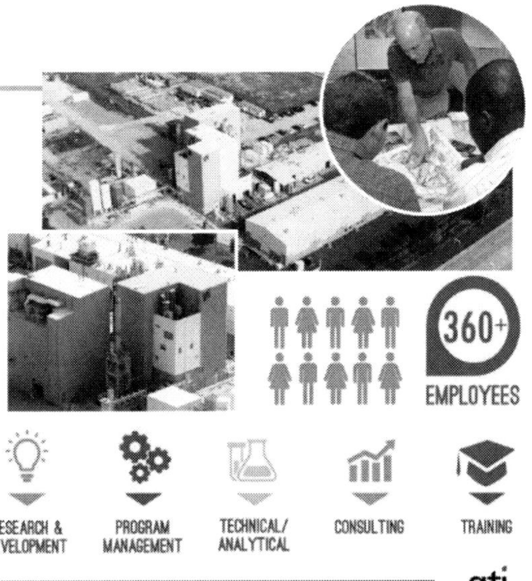

FOR A BETTER ENVIRONMENT AND A STRONGER ECONOMY

2 gti

Outline

- Cool GTL
  - What it is
  - What's unique and different
  - Applications and markets
  - Project status
- SoftOx
  - What it is
  - What's unique and different
  - Applications and markets
  - Project status

3 gti

Cool GTL

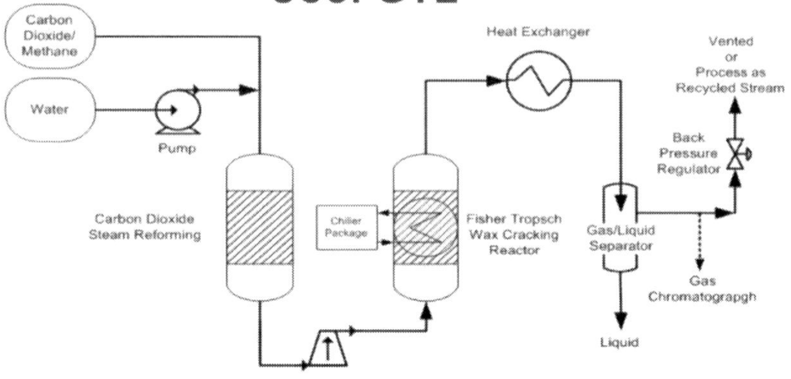

- Converts CO_2 rich methane, ethane and propane to high quality gasoline, diesel and jet Fuel
- Works well for any CO_2 or CO containing gas
- Uses Unique CO_2/steam reforming catalyst to directly make 2:1 H_2/CO synthesis gas
- Uses Fluid bed FT and wax cracking reactor
- Simple and compact/ unique catalysts in each stage

4 gti

Cool GTL Reactions

(I) $H_2O+CH_4 \rightarrow CO+3H_2$	CO and H_2 formation (800°C)	Reactor 1
(II) $CO_2+CH_4 \rightarrow 2CO+2H_2$	CO and H_2 formation (800°C)	Reactor 1
(III) $CO_2+H_2 \rightarrow H_2O+CO$	Water gas shift to equilibrium	Reactor 1
(IV) $CO+2H_2 \rightarrow -[CH_2]-+H_2O$	Hydro/oligomerization (200°C)	Reactor 2
(V) $H_2+-[CH_2]- \rightarrow -[CH_2]-+H_2$	Isomerization (200°C)	Reactor 2

5

gti

What's Unique and Different about Cool GTL ?

- Unique Catalyst in Cool Reforming Step
  - Robust with long life - minimal coking
  - Directly makes 2/1 H2/CO synthesis gas by adjusting amount of steam added
  - Simple and direct, mild temperatures. steady performance
- Unique Catalyst in Fischer Tropsch Step
  - No wax
  - Drop in gasoline, diesel and jet
  - Fluid bed process for easy heat removal

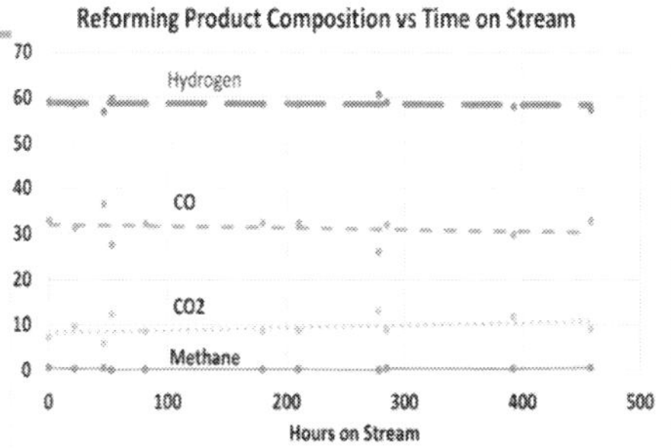

Clean Hydrocarbon Product

Low cost, simplified version of an old process

6

gti

Cool GTL was Designed from the Beginning for Modular Deployment

- Biogas and Oil Well Applications typically small in size
- Simplified process
- Small Footprint
- Compact

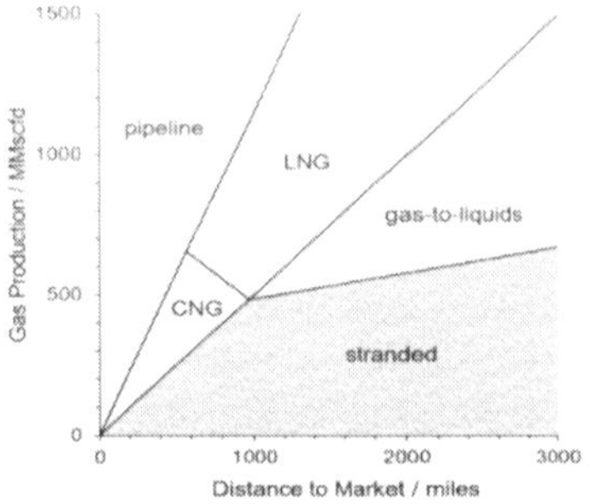

7 gti

Cool GTL Applications & Markets

- Biogas to GTL → Direct utilization of CO_2 methane, ethane, propane
- High CO_2 content natural gas → utilize some shut in natural gas
- Co-locate with acid gas plants → large market, economic synergies
- Flare gas mitigation → need CO_2, offshore, Africa

8 gti

IH$^{2®}$ Cool GTL Application

Case 1

BASE IH2 PROCESS
(showing more details)

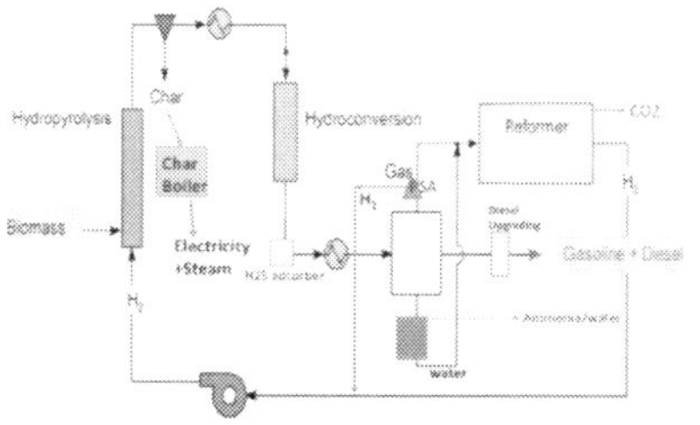

86 GPT Liquid Yield from wood

Case 2

IH2 plus CO2/H2O Reforming of IH2 Light Gas to Increase biogenic Liquid Yields to 38% - LCA- Case 2

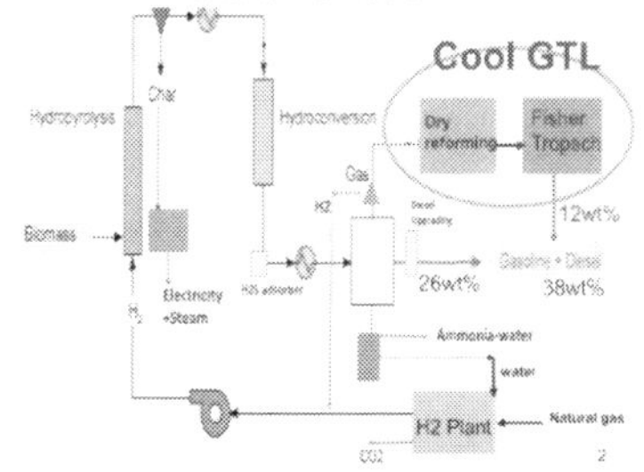

126 GPT Liquid Yield from wood

- Use natural gas to make H_2, use Biogenic gas to make more biogenic liquid
- LCA says IH^2+Cool GTL still above 60% GHG reduction

gti

Project Status

- **Two patent applications filed**
- **Requires longer term tests**
  - Demonstrate catalyst life
- **Requires larger scale skid mounted modular testing**
  - Demonstrate at larger scale
- **2 years from commercialization**
- **Requires a commercialization partner**

gti

SoftOx (Soft Oxidation)

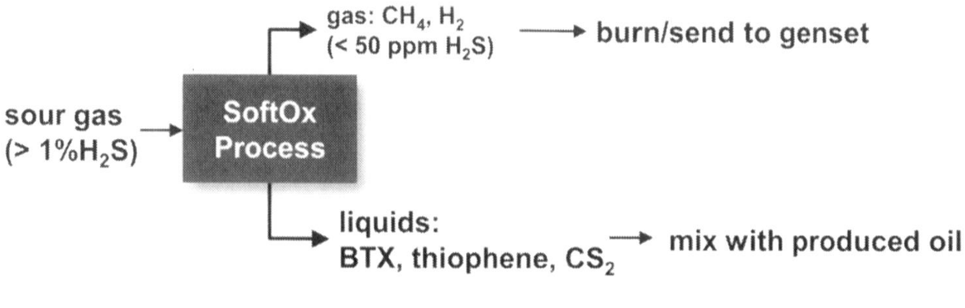

Convert wellhead sour gas into liquids that can be mixed with crude oil and a low-sulfur, H_2-rich gas for generating electrical power

11

Flaring of Sour Gas is a World Wide Problem

- Equipment Generating Electricity from natural gas typically requires low H_2S concentrations
- It's expensive to clean-up high H_2S levels in associated gas (gas associated with oil production)
- Therefore flaring of sour associated gas occurs which wastes energy and generates CO_2 and SO_2

SoftOx Solving the Sour Gas Issue

- Reducing H_2S to allow power generation
- Added liquids to production
- Modular design
- Remote well head applications

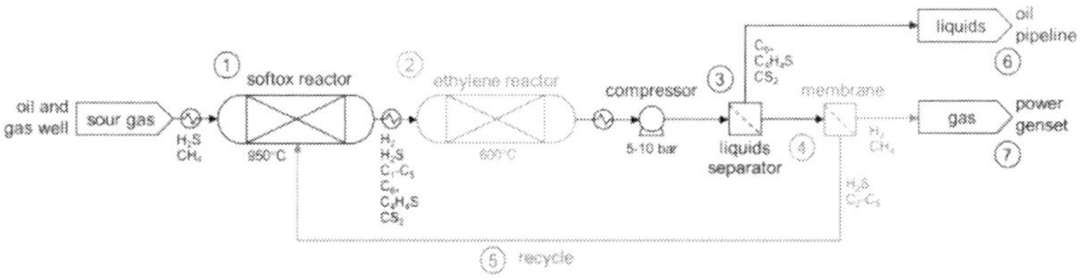

SoftOx Application Space

- Low flow, high H_2S
- Remote/stranded
- Little infrastructure
- Focused on flaring
- TRL = 4

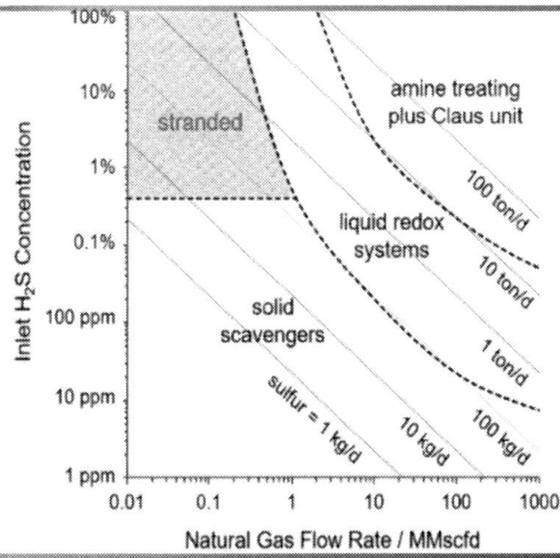

SoftOx Reactions

$CH_4 + 2H_2S \rightarrow CS_2 + 4H_2$	CS_2 formation	Reactor 1
$4CH_4 + H_2S \rightarrow C_4H_4S + 7H_2$	Thiophene formation	Reactor 1
$6CH_4 \rightarrow C_6H_6 + 9\ H_2$	Methane dehydroaromatization	Reactor 1
$3C_2H_6 \rightarrow C_6H_6 + 6H_2$	Ethane dehydroaromatization	Reactor 1
$C_2H_6 \rightarrow C_2H_4 + H_2$	Ethane to Ethylene	Reactor 1
$C_3H_8 \rightarrow CH_4 + C_2H_4$	Propane to $C_1 + C_2$	Reactor 1
$3C_2H_4 + H_2 \rightarrow C_6H_{14}$	Ethylene to C_4+	Reactor 2

15

Experimental Data - Conversion vs Temperature First and Second Reactor

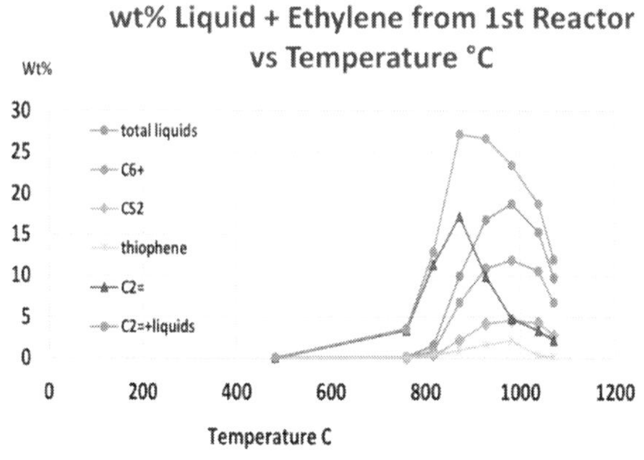

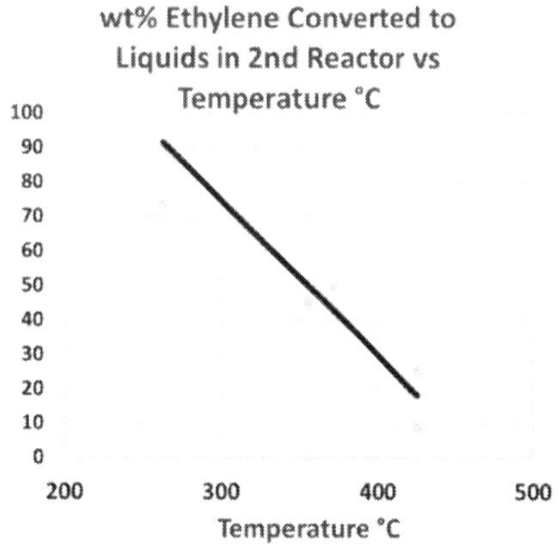

16

What's Unique and Different about SoftOx

- Totally unique concept for converting sour natural gas- avoids expensive H_2S cleanup
- Unique first stage catalyst with 28% conversion per pass to liquids and C_2=
- Unique second stage catalyst to oligomerize the dilute ethylene to make additional liquids- produces per pass liquid yield of 26%
- Unusual membrane to remove H_2S at low pressure and recycle it back

Totally new process concept

17 gti

SoftOx was Designed from the Beginning for Modular Deployment

- Oil Well Applications typically small in size 30,000 scfd-100,000scfd
- Simplified process
- Small Footprint
- Compact

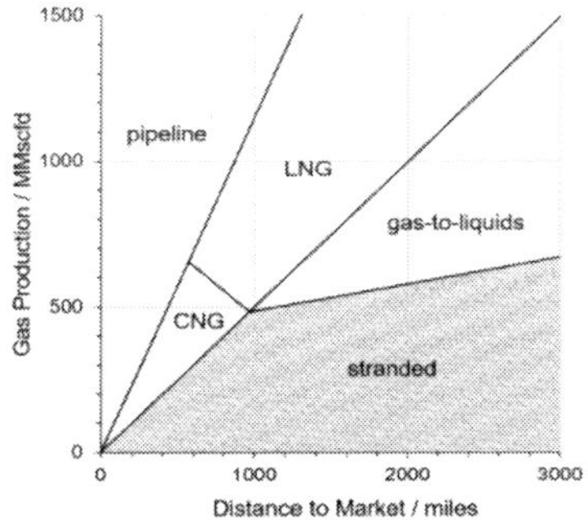

18 gti

SoftOx Status

- One granted patent- U.S. 9,963,399
- One application pending
- Requires longer term testing
- Requires integrated testing
- Requires larger scale skid mounted modular testing
- 2 years from commercialization
- Requires a commercialization partner/ funding

19 gti

Acknowledgements

- GTI Co-authors: Jim Wangerow, Senior Engineer, Dr. Pedro Ortiz-Toral , Manager
- Gas Technology Institute, Internal R&D funds
- Dr. Dane Boysen, Modular Chemical, for SoftOx Process
- US Department of Energy (Contract No. DE-0007009) for the Cool GTL work.
- US ARPA –E (Contract No- DE-AR-0000761) for SoftOx work

20 gti

Turning Ideas into Practical Solutions

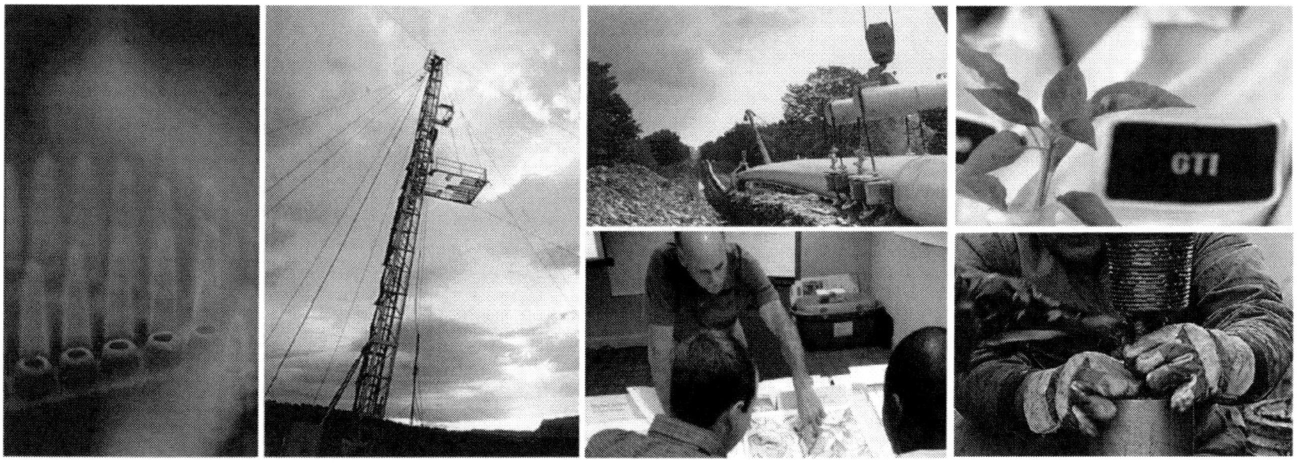

21

gti

Quick-fire Process Development Case Studies – Five Lessons Learned

Authors:

Santhany, S., The Dow Chemical Company

As a process research engineer, all chemical engineering disciplines merge upon us to make us most effective in our roles. This gives us opportunities to become deep technical experts narrow in certain fields of chemical engineering or to become 'expert generalists' that work across disciplines to bring it all together. In this talk, five lessons learned will be discussed in a quick fire manner that spans across Reaction Engineering, Solids Processing, Process Modeling, Economics, and Technology Transfer. With appropriate time, cost, and expertise everything can be done with a high level of success. Those three constraints are not always available to us in this fast pace world, which can lead to some lessons learned.

Development of Separation Processes and Applications using Hollow Fiber Membranes: Challenges and Opportunities

Authors:

Crowder, R., Pentair

Membrane systems can offer an effective alternative to more conventional unit operations. The fact that membranes work differently than older technologies can give them advantages in certain applications. Understanding how and when to use membranes will lead to more productivity, and fewer failures. To that end we will briefly look at three diverse membrane applications and understand the challenges and opportunities of each.

The first application is that of waste water treatment with an anaerobic membrane bioreactor (AnMBR). Membrane systems can handle higher strength waste water with a smaller footprint and a higher purity filtrate than conventional settling ponds. In addition, anaerobic MBRs produce a useful byproduct. The challenge is that sizing and pretreatment of the system must be done carefully to ensure a good outcome. In addition, some waste streams are more suited for an AnMBR than others.

The second application is that of beer filtration. Membranes provide benefits vs diatomaceous earth filtration in terms of cutoff, contamination, and scalability. Here the challenge is to understand the cleaning cycle and the transmembrane pressure during that cycle and operate these systems in a way to ensure longer periods of time between cleanings, leading to longer membrane life and higher productivity.

The third application is quite different from the first two. This involves a gas separation membrane used for drying compressed air or other gases. In this case the opportunity presented by membranes over pressure swing adsorption or refrigerated dryers is that of smaller footprint, simpler operation, and scalability for point of use. Membranes are not invulnerable, however, so care must be taken to avoid fouling or rupture. In this use, differences in operation and pretreatment can mean the difference between years of productive output and a short, marginally productive life.

CHEMISTRY THAT MATTERS™

CONTEXTUAL APPROACHES TO PROCESS DEVELOPMENT

DEVELOPMENT FOR PRODUCTS DEVELOPMENT, PROCESS IMPROVEMENTS, AND NEW PROCESS DEVELOPMENT

Ramesh Rameswaran
AIChE Process Development Symposium -2018
Oak Brook, IL

FROM A LARGE COMMODITY CHEMICAL COMPANY PERSPECTIVE

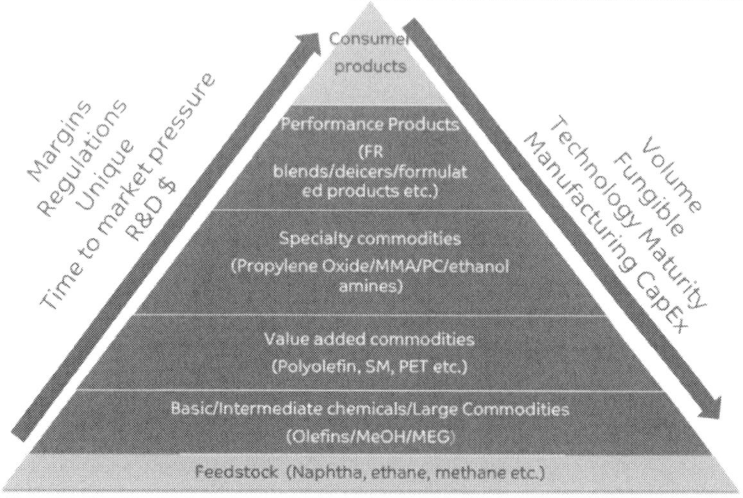

2

Classification: General Business Use

WHAT IS PROCESS DEVELOPMENT?

3

Classification: General Business Use

IDEAS ORIGINATE FROM DIFFERENT PLACES

Platform expansion

Reliability Improvement

Customer expectations

Anticipated applications

Abundance of raw material

Market trends

Reduced supply of raw material

Operational cost reduction

Safer routes

Value addition

Anticipated new regulations

Serendipity

4

IDEAS MAY FALL IN THREE BUCKETS

Abundance of raw material
Reduced supply of raw material
Safer routes

Reliability Improvement
Operational cost reduction
Improved safety

Platform expansion
Anticipated new regulations
Value addition

New Process Development | Process Improvements | New Product Development

DEVELOPMENT PROCESSES HAVE MANY COMMON THREADS

Fundamentals

- Profitability improvements
  - Capital efficiency
  - Raw material
    - Availability
    - Efficiency
  - Energy reduction
- Quality
  - Impurity specs
- Regulatory requirements
  - Environmental considerations
    - Emission
    - Waste disposal
    - Exposure
- Hazards
  - Process
  - Personal exposure

CONTEXTUAL APPROACHES TO PROCESS DEVELOPMENT

NEW PROCESS DEVELOPMENT

7

Classification: General Business Use

WHY WOULD ONE DEVELOP NEW PROCESS TECHNOLOGY?

Abundance of raw material

- Ube process for direct production of MEG from syn gas
  - Practiced in China
  - Cheap and abundant raw material – Coal
    - Lack of access to cheap ethane
- PDH – Availability of propane from shale gas and lack of it from naphtha crackers

Vagaries of Co-Products

- No co product process eg: Hydrogen Peroxide –Propylene to Propylene Oxide (HPPO)
  - Profitability of other processes depend use of chlorohydrin or co produces MTBE, Styrene monomer
  - Waste handling and cyclic nature of co-product demand

Demand by OEMs for low carbon footprint chemical production

- Bio based processes
- CO2 utilization
- Recycled plastic

8

TO DEVELOP A NEW PROCESS TECHNOLOGY...

Availability long term feedstock in various geographies needs evaluation

Corporate strategy for growth plans in various geographies needs to be considered

Identification of available technologies that matches various feedstock should be determined

- Identify gap in feedstock-technology match
  - Identify improvement opportunities
  - Develop R&D strategies for providing business with options

A PROCESS TECHNOLOGY DEVELOPMENT TAKES TIME (>10+ Years) AND MONEY (>$40Million) To DEVELOP

COST IS THE PRIMARY FOCUS IN DEVELOPING A PROCESS TECHNOLOGY

For commodity chemical process technology $250 to >$1,000 million capex can be expected

When the proof of chemistry concept is demonstrated

- Evaluate preliminary capex estimate is made
- Compare capex with incumbent technologies
- Compare raw material margin vs incumbent processes
- If sufficient delta is anticipated full scale development program may be commissioned

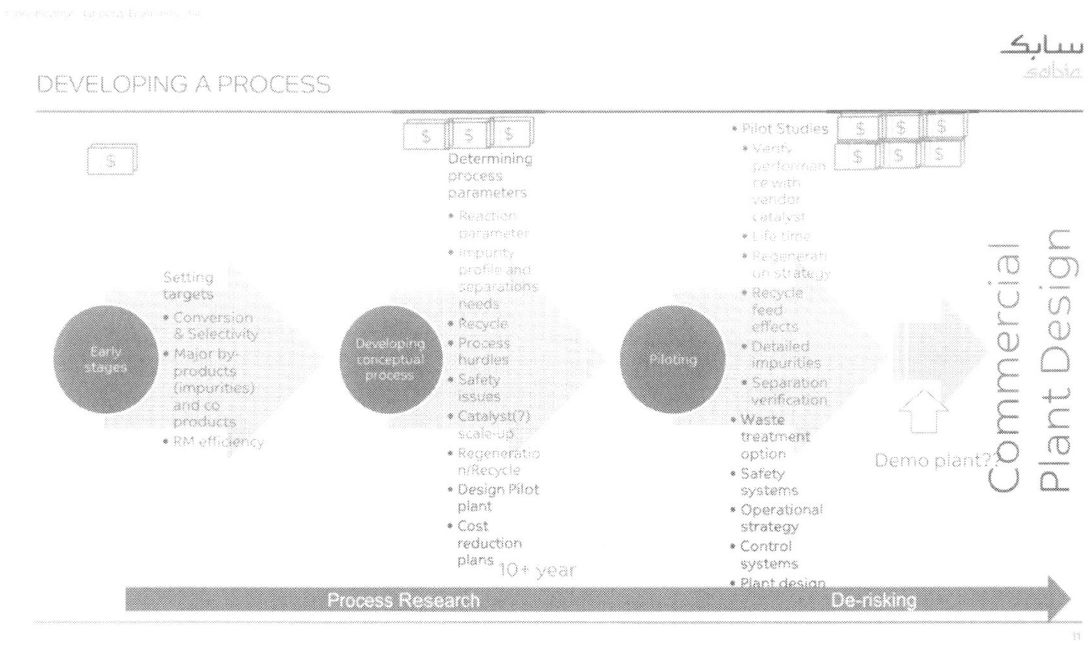
سابك
sabic
DEVELOPING A PROCESS
$
$ $ $
Early stages
Setting targets
• Conversion & Selectivity
• Major by-products (impurities) and co products
• RM efficiency
Developing conceptual process
Determining process parameters
• Reaction parameter
• Impurity profile and separations needs
• Recycle
• Process hurdles
• Safety issues
• Catalyst(?) scale-up
• Regeneration/Recycle
• Design Pilot plant
• Cost reduction plans
10+ year
Piloting
• Pilot Studies
• Verify performance with vendor catalyst
• Life time
• Regeneration strategy
• Recycle feed effects
• Detailed impurities
• Separation verification
• Waste treatment option
• Safety systems
• Operational strategy
• Control systems
• Plant design
$ $ $
$ $ $
Demo plant??
Commercial Plant Design
Process Research
De-risking
11

CONTEXTUAL APPROACHES TO PROCESS DEVELOPMENT
DEVELOPMENT FOR PROCESS IMPROVEMENT

PROCESS IMPROVEMENT IS INCREMENTAL IN NATURE - 1

Normally within the confines of operating plants and designs
Changes involves only small sections of the plant and not the whole plant
In general, low capital

Reliability improvements

- Equipment failure
  - Maintenance record
    - Detailed diagnostics

} Most studied and evaluated

- Sudden changes in processability (Example: Fouling heat exchanger, foaming, sticking valves, polymer buildup)
  - Evaluating upstream process condition changes
    - Changes in chemistry
      - Trace chemistry issues are difficult to identify
      - Lab simulation might be needed
  - Feedstock spec changes

13

PROCESS IMPROVEMENT IS INCREMENTAL IN NATURE - 2

Quality Improvements

- Post treatment options to remove trace impurities for specific customer need
  - Safety consideration
  - Residue removal
  - Toll or not to Toll this treatment

Reducing yield losses

- Evaluation of product losses and locations
- Recovery options for lost products
  - waste stream evaluation

Improving reaction yield

- Reactor condition changes
  - Detailed modeling and verification in the plant (plant trials with process/safety limits of equipment)
  - Process Scale down to study the process conditions
- Catalyst
  - Drop in
  - Long qualification process (2+ years)
  - New impurities???

14

DROP-IN CATALYST DEVELOPMENT

Identify need and expected performance characteristics
Look for existing vendor of similar catalysts

- If the catalyst vendor supplies catalyst for the same service
  - Compare performance
  - Look for reference
- If the catalyst is used for other services, screen and select catalyst
  - Go through full battery of evaluation in bench scale reactors
    - Deactivation and regeneration cycles
    - Yields (Specifically, conversion and selectivity) at allowable operating limits at the plant
    - Impurities
    - Initiate
- If no catalyst is available, develop in-house catalyst
  - Screening
  - Scale up at a catalyst manufacturer
  - Study
- Plant trials

DROP IN CATALYSTS

	Custom in-house	Vendor supplied
Lab testing	√	√
Scale up at catalyst manufacturer	√	
Pilot testing for yield improvements and life (in-house or at vendor)	√	√
Plant testing	√	√

Multiple years to qualify a catalyst as drop in catalyst

REPURPOSING EXISTING CAPITAL EQUIPMENT
(OF COURSE WITH MINIMUM OR NO ADDITIONAL CAPITAL)

Unit shutdown for many reasons

- Regulation changes
- Business strategy changes
- Not meeting the market needs
- Consolidation into larger assets
- Material that are made in campaigns
  - Fill-in idle period

Challenges

- Square peg into round hole
- Equipment in wrong sequence

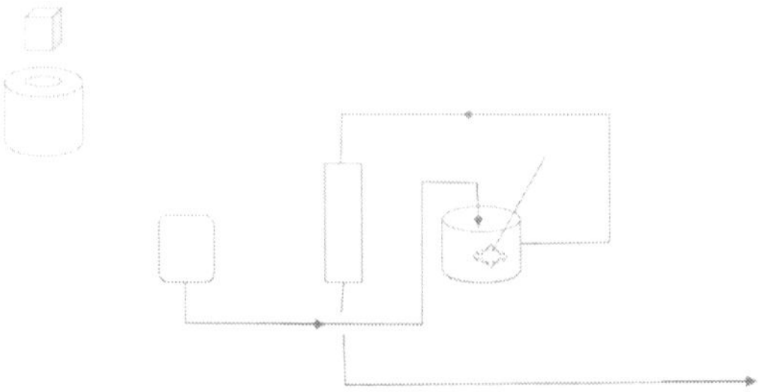

PROCESS DEVELOPMENT FOR NEW PRODUCTS

NEW PERFORMANCE PRODUCTS

High margin materials

Low volume

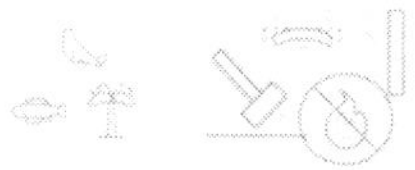

Short time to market

Performance paramount
- Formulating to obtain desired properties that are critical for performance takes time
- Can afford some raw material inefficiencies
- In general, very tight specification

CapEx varies with product
- Platform expansion – no change or minor changes to existing equipment
  - Emphasis is then not to lose the functionality of the original design
- New Product – Toll manufacturer, Repurpose equipment, new build
  - Process development for tech transfer (to fit Toller facility)

19

PRODUCT DEVELOPMENT PROCESS

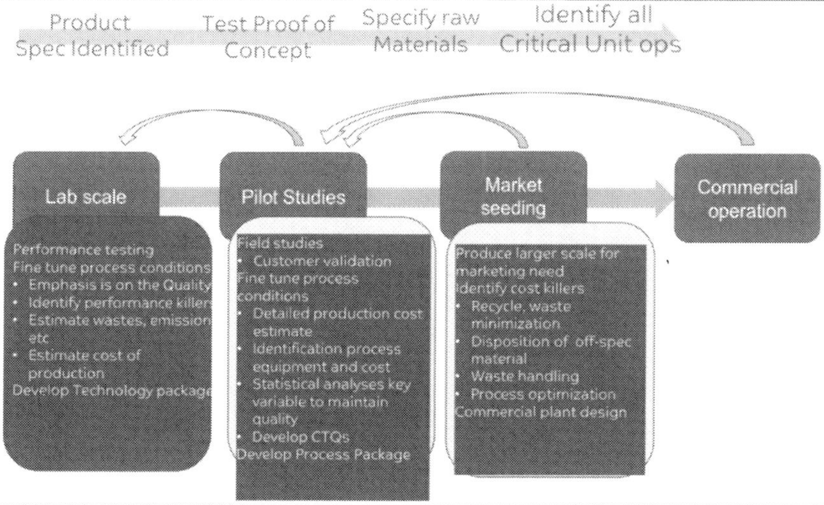

20

ROLE OF PROCESS DEVELOPMENT

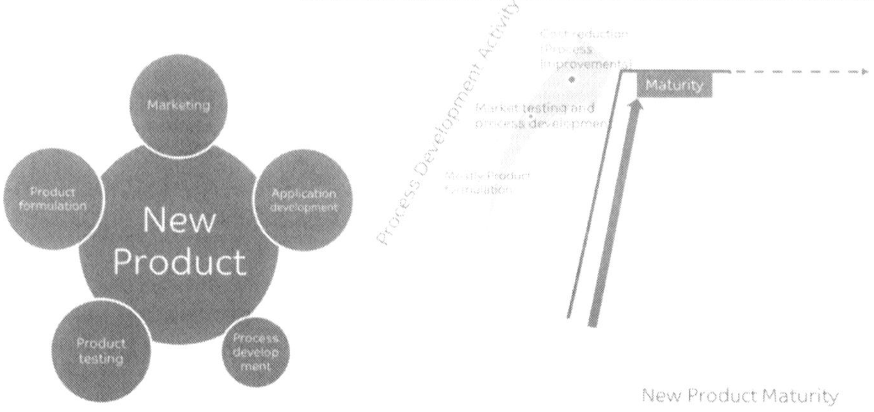

THINGS TO KEEP IN MIND

New products require permits & notifications from various governmental authorities

If new products are for seasonal applications field test time is limited

- Example : aircraft deicers, agricultural products, etc

Products developed for one geography may not be applicable in other parts of the globe

SUMMARY

	New Process	Process Improvement	Product introduction
Motive	• Disruptive technology • Advantaged Feedstock • Regulatory issues • Avoidance of co- & by products	• Improve safety • Improve profitability	• Need for lower cost performance material • Regulatory compliance
Process Development	All the unit ops needs work	Specific areas	Emphasis on performance then Process development
CAPEX	High	Low	Depends?
Product performance/Quality	Meet or beat completion	No changes at minimum	Totally on performance and quality
Incumbent technology	Mature	New replacement may be available	May be new
Time to market	~10 year	When possible	Short (1-5 yrs)
Risk	High	Low	Depends ??
Rewards	High	Incremental	High

23

Process Development of Opteon® yf – Innovating to Meet Market Challenges

Authors:

Krause, K., The Chemours Company

Regulatory response to climate change, first enacted in Europe, required development of a low global warming potential (GWP) refrigerant for vehicles on an accelerated timetable. Innovative approaches to identify a drop-in replacement for the legacy refrigerant, achieve market acceptance, develop a process for first-to-market commercial production, and develop an improved process for larger scale production will be described in this poster. Key aspects include parallel process development programs, partnerships for process development and commercialization, strong intellectual property protection, leveraging of existing processes coupled with development of new technology, and process simplification.

Enabling Rapid Process Development for a Challenging Salt Crystallization

Authors:

Mohan, A., Merck & Co.

The final chemical transformation and isolation in the synthesis of an Active Pharmaceutical Ingredient (API), referred to as the Pure Step, is often chemically simple but scientifically, operationally and strategically the most challenging. Pure step development is critical because it is used to determine and support the critical quality attributes (CQAs) for the API, which will have lasting impacts on the process. This poster will detail specific challenges for an API that is isolated as a citrate salt, crystallized out of methanol and isopropanol. This citrate salt is then formulated via direct compression to make the deliverable for the patient. The salt crystallization is particularly challenging due to: (1) the propensity of the citrate salt crystal to form solvates, (2) the particle size control requirements, and (3) the variability in starting material purity. The team has had to execute targeted, rapid process development in order to support pilot plant API batches which will supply clinical trials. This work has required not just the technical capabilities of the team, but has also required streamlined collaboration and flawless communication across both drug substance and drug product space. This poster will detail the comprehensive and hefty process development work which was executed to support the Pure Step, focusing on the drug substance crystallization, and the impacts on the drug product space.

Scale Up of Fluidized Bed Reactors: What Can Be Calculated and What Needs to Be Measured

Authors:

Freireich, B., Particulate Solid Research, Inc.

Fluidized beds are commonly used for the manufacturing of a wide range of products including syngas, fuels including gasoline, acrylonitrile, polyolefins, and ultra-high purity silicon. These units provide unsurpassed heat transfer along with the ability to flow solids during operation. However, scale-up of these units can be challenging with respect to productivity and reliability. For fast reactions, the fluidized bed is usually mass-transfer limited and mass-transfer is strongly dependent on bed hydrodynamics. Thus, scale-up parameters need to include gas compression, bubble growth, solids mixing, and particle cohesive forces all of which need to be obtained on a large-scale. Many of the available correlations were either obtained on too small of a test unit to be relevant with commercial application, or simply don't consider parameters that affect to mass transfer.

Thus, understanding key hydrodynamic parameters such as bed density profiles, bubble hydrodynamics, entrainment, pressure loop profiles and jet-penetration lengths, all of which are depend on particle properties, can yield a commercial-scale design that mimics pilot-plant performances with reliability goals similar to other unit operations. While modeling efforts can provide good approximations of what is and is not important with scale up. Yet, most models are based on "ideal" or "standard" conditions and properties and rarely consider micro-scale effects such as interparticle forces (coulombic, capillary, van der Waals) or other surface interactions (collisional stresses). Thus, experiments from cold flow to pilot studies are needed to understand controlling hydrodynamic parameters as relevant to the operation conditions of the process and the physical properties of the powder or catalyst.

Selection of Isolation Methods for Micron Sized Particles

Authors:

Kumbhalkar, M., The Dow Chemical Company

Solid-liquid separations are commonly used in industrial processes and are approached in a distinctive way based on the desired component- solid or liquid. The separation technique is dependent on a combination of factors- how the feed suspension was prepared, total volume, solid concentration, and particle shape and size distribution of the solids. Investigating the performance of various separation equipment is time consuming and this search can be narrowed down by utilizing specific laboratory scale tests to predict the behavior of a separation technique before scale-up. Herein, the processes of technology selection for situations where solids are the desired fraction with particle size < 10 μm are discussed. The choice of cake filtration methods such as vacuum and pressure filtration is often based on the average flux, which indicates the time required for filtration. Filter media with small channel sizes are required to isolate the micron sized particles. The cake deposited always exhibits high resistance and the small particles may block the channels leading to unacceptably long filtering time. Sedimenting centrifuges such as tubular, or stacked disc can be used for isolating sub-micron particles since they do not use a filter medium for separation and avoid the challenges of media fouling. The selection of sedimenting centrifuge can be based on laboratory scale feasibility tests performed with a benchtop centrifuge by varying the centrifugal force and residence time and observing the sediment compactness and supernatant clarity. Other filtration methods including thin cake filtration and cross-flow filtration can also be considered for systems with small particles.

Design and Characterization of a Lab Scale Pump Loop System

Authors:

Larimer, J., The Dow Chemical Company

Turbulent immiscible liquid-liquid systems produce functionalized highly branched polymers via hydrolysis and condensation reactions. Due to simultaneous and interdependent physics, including coalescence, dispersion, heat and mass transfer, and competitive consecutive chemical reaction, time dependent mixing characteristics impact the resulting material produced.

Well characterized mixing equipment such as stirred tank reactors have well established rules for designing and scaling mixing dependent processes. Specific guidelines exist for both geometric and operational parameters such as load volumes, tank and impeller diameters, impeller types, number and location, and reactant addition strategies. Pump loop systems, however, have not been explicitly studied in the literature; therefore, scaling guidance for these systems is limited. Parallels can be drawn between well characterized stirred tank systems and pump loop operation, however, there are more independent design and operational variables in pump loop systems that increase the level of difficulty in process scaling over that of the stirred tank design.

Additionally, highly branched polymers can grow in three dimensional space as a function of time. Reaction conditions as a function of time influence how the structure of these species build[1]. Different mixing conditions throughout different zones of a pump loop system can shift the instantaneous growth of a molecule between different reaction mechanisms. Therefore understanding what aspects of pump loop system design have the largest influence in how molecules build is essential to successfully scaling a process. This poster highlights the challenges associated with designing and tuning a pump loop process asset.

REFERENCES

1. Brinker, C.J., "Hydrolysis and Condensation of Silicates: Effects on Structure", Journal of Non-Crystalline Solids, V100, 1988, Pg 31-50, North Holland, Amsterdam. Elsevier Science Publishers B.V.

Modular Chemical Process Intensification (MCPI): A Perspective from the RAPID Manufacturing Institute

Authors:

Palou-Rivera, I., RAPID

In December 2016, the Department of Energy announced the establishment of the 10th Manufacturing USA Institute, representing a critical step in the federal government's effort to double U.S. energy productivity by 2030. The Rapid Advancement in Process Intensification Deployment (RAPID) Institute (https://www.aiche.org/rapid) is focused on addressing barriers to enable the development of breakthrough technologies to boost energy productivity and energy efficiency through manufacturing processes in industries such oil and gas, pulp and paper and various domestic chemical manufacturers. RAPID will leverage approaches to modular chemical process intensification (MCPI) — such as combining multiple process steps as mixing, reaction, and separation into single more complex and intensified processes — with the goal of improving productivity and efficiency, cutting operating costs, and reducing waste.

This poster will present the approach used by the RAPID Manufacturing Institute to advance MCPI in the selected Focus Areas of Chemical and Commodity Processing, Renewable Bio Products, Natural Gas Upgrading, Modeling and Simulation, Intensified Process Fundamentals, and Module Manufacturing. Achievements to date such as the compile roadmap and plans for 2018 and beyond will be the cornerstones of the presented material.

Polysaccharide and fructose syrup production using GTF enzymes

Kevin D. Nagy, DuPont Industrial Biosciences

Commercialization of chemicals and materials derived from bio-based sources requires finding opportunities that have the right balance between performance, cost, and sustainability. A significant challenge using fermentation technology is the loss of carbon to CO_2; carbon yields can be significantly lower than 50%, and these low yields have substantial implications on both the process economics and environmental footprint. One approach to improve these metrics is to use enzymatic processes instead of microbial fermentations. The production of engineered polysaccharides using glucosyltransferase (GTF) enzymes is an example of a technology that allows for 100% of the carbon to be recovered as valuable products. These enzymes convert sucrose into a polysaccharide comprising glucose while simultaneously generating an enriched fructose stream. This method of producing fructose is also advantaged compared to the typical process using a starch feedstock. Commercial fructose syrup products contain 42-55 wt% dry weight basis (DWB) fructose. The 55 wt% DWB fructose stream cannot be produced directly and requires several chromatographic steps to achieve an intermediate stream containing 90+ wt% DWB fructose. This high purity fructose stream is then back blended to achieve the right composition. Overall, this incumbent process is extremely energy intensive due to the large amount of water evaporated in the chromatography process. Comparatively, the glucosyltransferase approach outlined herein directly generates a high purity fructose syrup containing 75+ wt% DWB fructose suitable for blending while completely avoiding chromatography and the associated evaporation requirements. These glucosyltransferase based processes thus enable the production of new polysaccharide materials while simultaneously streamlining the production of fructose syrup.

Developing new bio-based chemicals and materials requires balancing several considerations

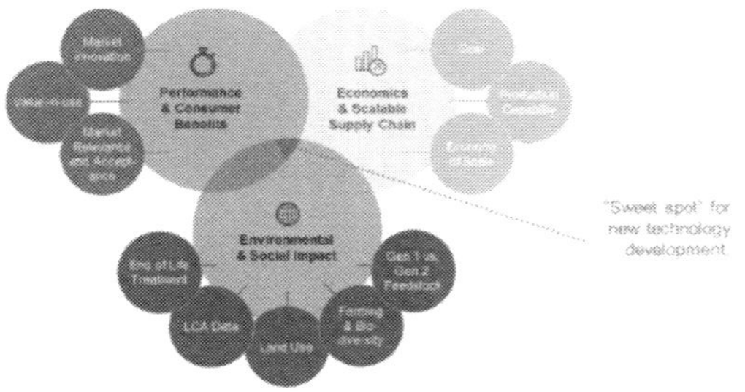

Enzymatic polymerization is a high carbon yield process that yields two valuable products

Yields of industrial bioprocesses

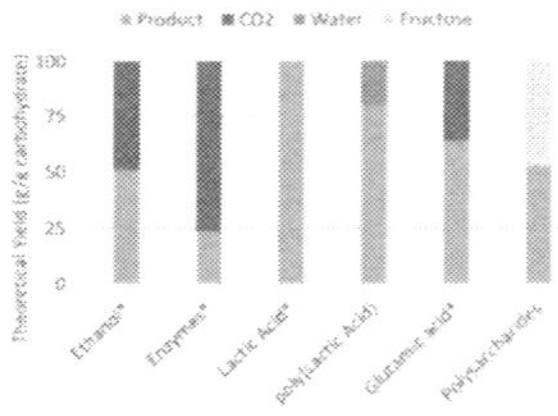

*Industrial Crops and Products 20 (2004) 75-88

Products from enzymatic polymerization

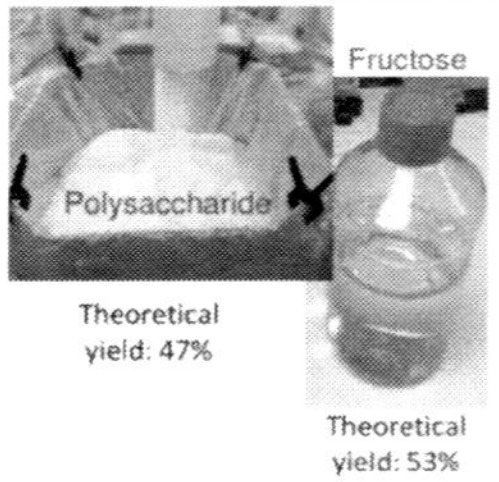

Sugar beet and sugar cane enable exceptional land use efficiency

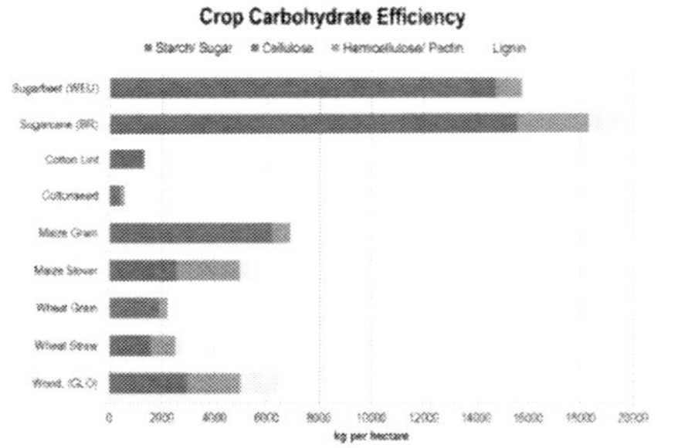

Engineered polysaccharides have broad utility across a wide range of applications

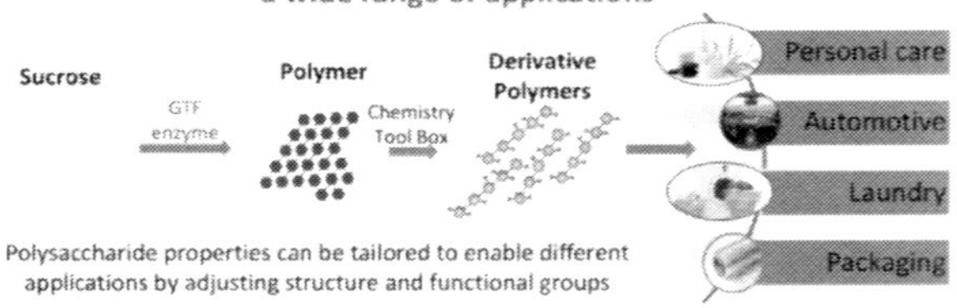

Polysaccharide properties can be tailored to enable different applications by adjusting structure and functional groups

Glucosyltransferase (GTF) enzymes convert sucrose into polysaccharides and soluble sugars

GTF from *S. mutans*

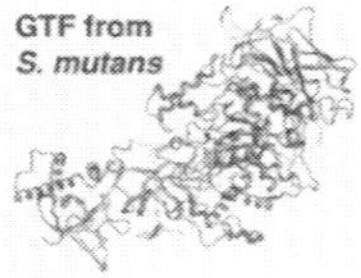

- Glycoside-Hydrolase family 70 (GH70)
- Monomer with typical MW of 160 kDa
- Extracellular enzymes, mainly produced by:
  - Streptococcus species
  - Leuconostoc species
  - Lactobacillus species

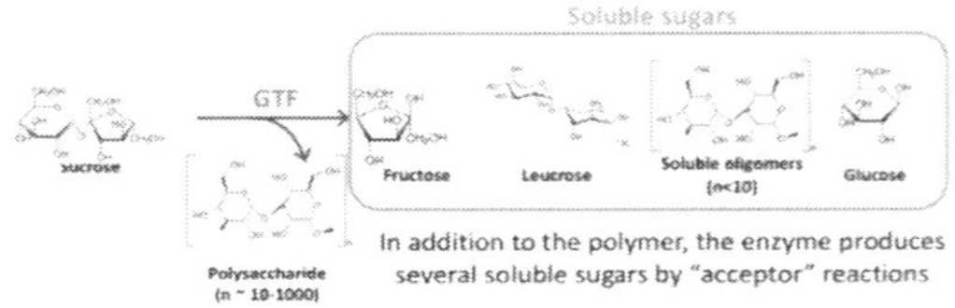

In addition to the polymer, the enzyme produces several soluble sugars by "acceptor" reactions

Protein engineering has increased both the polymer yield and fructose stream purity

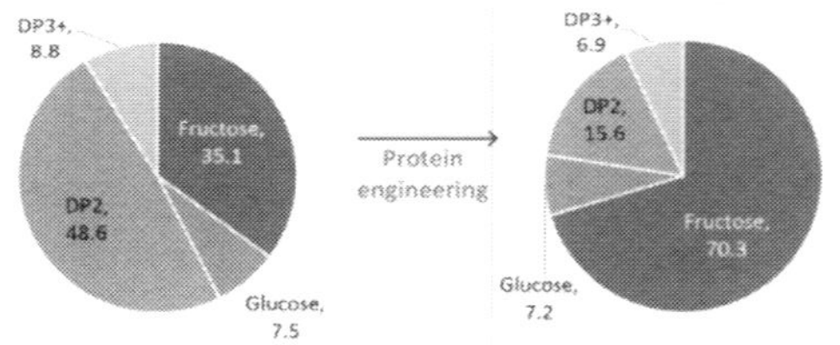

F…osition using …enzyme

Fructose composition using improved GTF enzyme

Byproduct sugars can be eliminated using hydrolysis or nanofiltration technologies

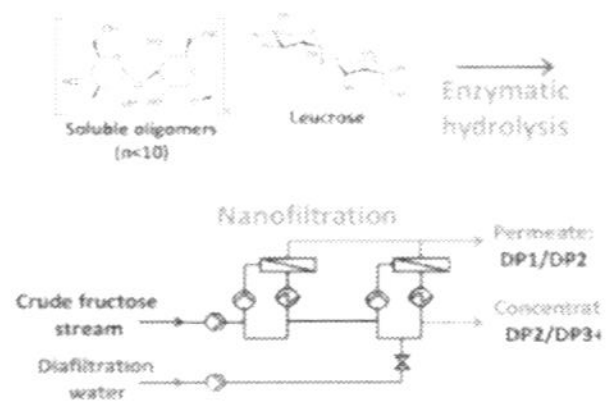

A highly purified fructose stream can be achieved through further processing

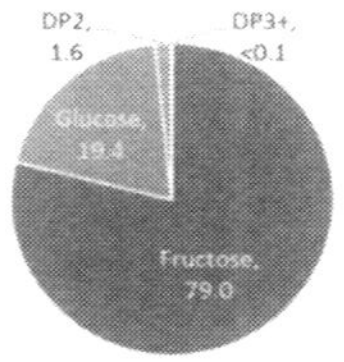

Upgraded fructose composition

Manufacture of sweeteners

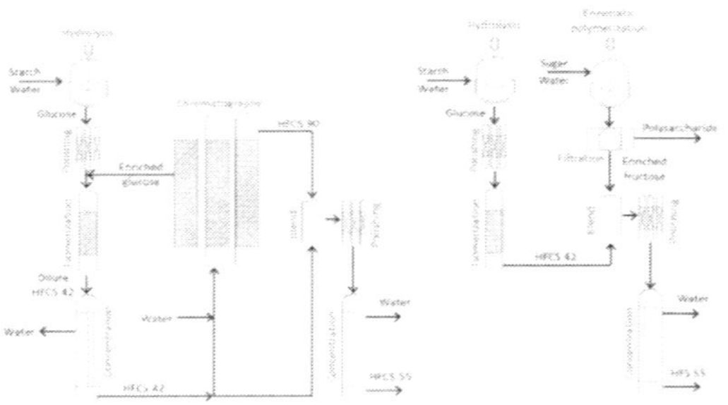

Production of sweeteners from corn requires several energy intensive processing step

Sweetener production can be simplified using enzymatic polymerization technology

Conclusions

The enzymatic conversion of sucrose into polysaccharides and fructose syrup using glucosyltransferase enzymes can be made into a high yield process. The selectivity of the biocatalyst used in this process has been improved using protein engineering technologies, and the properties of polysaccharides produced using this technology can be tailored to enable applications in a broad array of industries. Furthermore, the soluble sugar produced using this process can be upgraded to yield a highly purified fructose stream. This stream can be integrated into and enables process simplification in large volume sweetener applications such as fructose syrup.

DuPont Industrial Biosciences

Simulated Moving Bed Reactor for Enhancing the Productivities of Equilibrium Limited Reactions

Authors:

Sreedhar, B., The Dow Chemical Company

Equilibrium-limited reactions are ubiquitous in the chemical industry. Those with highly unfavorable equilibrium constants face very energy intensive operations for conversion and recovery of the product, often needing extensive recycle. Chances are also that such severely equilibrium-limited reactions using conventional reactors are written off at an early stage and are never scaled up to production.

In this work, we demonstrate the potential of simulated moving bed reactor (SMBR) in increasing the productivity of strongly equilibrium-limited reactions and possibly rendering them economically feasible. In an SMBR, in-situ chromatographic separation of the products aids in shifting the equilibrium, thereby improving the conversion.

Liquid phase transesterification of propylene glycol methyl ether (DOWANOL™ PM) to propylene glycol methyl ether acetate (DOWANOL™ PMA) using a homogeneous catalyst is considered as a case study to compare the performance of SMBR with that of a conventional batch reactor. Here, DOWANOL™ PM is used both as a reactant and desorbent for the chromatographic separation inside the SMBR. In the SMBR process, an anion exchange resin, AMBERLITE™ IRA904 in chloride form is used as a non-catalytic adsorbent, while sodium alkoxide of PM is supplied as a homogeneous catalyst in the desorbent. Using model based optimization, we demonstrate the advantage of employing a non-catalytic adsorbent in the SMBR for the homogeneously catalyzed reaction. With the ability to independently tune the separation of the product in the SMBR, it is demonstrated that the productivity of the process increases with increasing separation factor. Presence of an optimal product separation factor is observed with respect to the amount of DOWANOL™ PM used for reaction and separation. At a fixed conversion of 95%, SMBR outperforms a batch reactor by increasing the productivity by 85% and reducing DOWANOL™ PM usage to 50% of what is required by the batch reactor.

Chemicals from Biomass: The Design of a Succinic Acid Production Facility

Authors:

Ozturk, S., Milwaukee School of Engineering

Using SuperPro Designer, a process simulation software, our team explored a biochemical route to producing succinic acid that uses the rumen bacteria Actinobacillus succinogenes to ferment corn stover. Key steps of the facility include feedstock pretreatment, bacterial fermentation, recovery of the final product by crystallization, and power generation.

The model was used to assess the quality of the theoretical product and the economic and environmental impact of the facility to ultimately determine if the design could compete with both the petrochemical process and existing bio-based solutions. The team's design is on par with the current market by producing 11.55 kilotons of 99.7% pure succinic acid annually, and offering a payback time under 5 years. Our team created a biorefinery design that maximizes profitability for a feasible and sustainable solution.

AUTHOR INDEX

AIChE
120 Wall Street, FL 23
New York, NY 10005-4020

ISBN 978-1-5108-7280-6

CURRAN ASSOCIATES INC.
proceedings.com